Lecture Notes in Biomathematics

Managing Editor: S. Levin

58

Catherine A. Macken
Alan S. Perelson

Branching Processes
Applied to Cell Surface
Aggregation Phenomena

Springer-Verlag
Berlin Heidelberg GmbH

Editorial Board

J. D. Cowan W. Hirsch S. Karlin J. B. Keller M. Kimura
S. Levin (Managing Editor) R. C. Lewontin R. May J. D. Murray G. F. Oster
A. S. Perelson T. Poggio L. A. Segel

Autors

Catherine A. Macken
Centre for Computing and Biometrics, Lincoln College
Canterbury, New Zealand

Alan S. Perelson
Theoretical Division, University of California
Los Alamos National Laboratory
Los Alamos, NM 87545, USA

Mathematics Subject Classification (1980): 60J85, 60K35, 92-02, 92A09, 80A30

ISBN 978-3-540-15656-7   ISBN 978-3-642-52115-7 (eBook)
DOI 10.1007/978-3-642-52115-7

This work is subject to copyright. All rights are reserved, whether the whole or part of the material is concerned, specifically those of translation, reprinting, re-use of illustrations, broadcasting, reproduction by photocopying machine or similar means, and storage in data banks. Under § 54 of the German Copyright Law where copies are made for other than private use, a fee is payable to "Verwertungsgesellschaft Wort", Munich.

© by Springer-Verlag Berlin Heidelberg 1985
Originally published by Springer-Verlag Berlin Heidelberg New York Tokyo in 1985

2146/3140-543210

PREFACE

Aggregation processes are studied within a number of different fields--colloid chemistry, atmospheric physics, astrophysics, polymer science, and biology, to name only a few. Aggregation processes involve monomer units (e.g., biological cells, liquid or colloidal droplets, latex beads, molecules, or even stars) that join together to form polymers or aggregates. A quantitative theory of aggregation was first formulated in 1916 by Smoluchowski who proposed that the time evolution of the aggregate size distribution is governed by the infinite system of differential equations:

$$\frac{dc_k}{dt} = \frac{1}{2} \sum_{i+j=k} K_{ij} c_i c_j - c_k \sum_{j=1}^{\infty} K_{kj} c_j \quad , \quad k = 1, 2, \ldots \quad , \tag{1}$$

where c_k is the concentration of k-mers, and aggregates are assumed to form by irreversible condensation reactions [i-mer + j-mer → (i+j)-mer]. When the kernel K_{ij} can be represented by $A + B(i+j) + Cij$, with A, B, and C constant; and the initial condition is chosen to correspond to a monodisperse solution (i.e., $c_1(0) = c_0$, a constant; and $c_k(0) = 0$, $k > 1$), then the Smoluchowski equation can be solved exactly (Trubnikov, 1971; Drake, 1972; Ernst, Hendriks, and Ziff, 1982; Dongen and Ernst, 1983; Spouge, 1983; Ziff, 1984). For arbitrary K_{ij}, the solution is not known and in some cases may not even exist. In chemistry and biology, aggregation processes generally are more complex than those described by the Smoluchowski equation. The process may be reversible and thus fragmentation terms need to be added (cf. Dongen and Ernst, 1983, 1984). Equation (1) then becomes

$$\frac{dc_k}{dt} = \frac{1}{2} \sum_{i+j=k} (K_{ij} c_i c_j - F_{ij} c_{i+j}) - \sum_{j=1}^{\infty} (K_{kj} c_k c_j - F_{kj} c_{k+j}) \quad . \tag{2}$$

Determining the appropriate form for the aggregation kernel, K_{ij}, that describes the rate of formation of an (i+j)-mer from an i-mer and a j-mer, and the fragmentation kernel, F_{ij}, that describes the rate at which the reverse process occurs, involves solving various combinatorial problems (e.g., determining the number of ways a k-mer can be formed by association of i-mers and j-mers, i + j = k, or by dissociation of n-mers, n > k). Moreover, if the reactions involve more than one type of monomer, as in the case of antigen-antibody or ligand-receptor interactions, then further complications arise in determining the appropriate aggregation and fragmentation kernels. Realistic models in chemistry and biology thus tend to give rise to very complicated versions of the Smoluchowski equation (cf. Samsel and Perelson, 1982, 1984; Perelson and DeLisi, 1975). In recent years, an enormous amount of progress has been made in studying these complex aggregation processes. Much of this modern work has not yet found application in the biological literature. Here we shall formulate what we believe to be a very important class of biological aggregation problems and solve them using one of the more powerful modern techniques: the theory of branching processes.

The question that we shall consider arises in studies of the interaction of cells with their environment. Cells have on their surfaces receptors for certain solution phase molecules, generally called ligands. The binding of a ligand to its complementary receptor is, in many cases, not the signal that the cell responds to. Rather, for reasons to be discussed later, the cell appears to respond only to the formation of receptor aggregates.

A long-standing problem in theoretical cell biology has been to predict the distribution of sizes of aggregates that form on cell surfaces when multivalent ligands bind and cross-link multivalent receptors. Here we take a very general approach and demonstrate the use of the mathematical machinery of branching processes in solving this problem. The basic method, due to Gordon (1962) and Good (1963), for using branching processes to compute the number of ways an aggregate can form, is reviewed and presented in the framework of receptor-ligand and antibody-antigen interactions. By way of example, we use the branching process method to compute the distribution of aggregate sizes produced in the random self-aggregation of f-valent molecules (thus reproducing the famous Flory-Stockmayer result), and the distribution of aggregate sizes produced in antigen-antibody reactions in solution (thus reproducing a result due to Goldberg, 1952). By incorporating the ligand binding model developed by Perelson and DeLisi (1980), we then compute the size distribution of aggregates formed when bivalent ligands interact with bivalent cell surface receptors. When the interacting monomers are both bivalent, aggregates are either linear chains or rings, and simpler methods can be used to determine the aggregate size distribution. In fact, our results agree with previous calculations of Dembo and Goldstein (1978) and Perelson and DeLisi (1980). However, when either the ligand or receptor has more than two reactive sites, branched networks can form on the cell surface. Computing the size of such networks has been a difficult, unsolved problem. Here we use the branching process method to solve this problem in general: we find the size distribution of aggregates formed when f-valent ligands, $f \geq 1$, bind and cross-link g-valent receptors, $g \geq 1$.

In idealized systems, an infinite receptor-ligand complex can form. We derive a necessary condition for the formation of such an infinite-sized aggregate or "gel" and then show how one can compute the distribution of sizes of aggregates belonging to the "sol" phase. We also show that inconsistencies discovered in previous models of receptor cross-linking by multivalent ligands are due to neglecting the distinction between cell surface "sol" and "gel" phases.

The branching process method is powerful in its generality. Its application is not limited to receptor-ligand problems but is appropriate to problems in many areas of biology. We believe that the techniques we develop in these notes will be valuable in increasing the accessibility of the branching process method as well as providing some important new results for cell-surface aggregation phenomena.

We thank Carla Wofsy for critically reading the manuscript. This work was performed under the auspices of the U.S. Department of Energy. C.A.M. was supported by N.I.H. grant 5 T32 GM07661 and would like to thank the members of the

Theoretical Biology and Biophysics Group at Los Alamos National Laboratory for their hospitality during her many visits. A.S.P. was the recipient of an N.I.H. Research Career Development Award AI00450 during the period in which these notes were written.

                                                    Catherine A. Macken

                                                    Alan S. Perelson

                                                    December 1984

TABLE OF CONTENTS

Preface

1. Introduction 1
   A. Receptors, Ligands, and the Importance of Receptor Clustering
      in Transmembrane Signaling 1
   B. Quantitative Theories of Receptor Clustering 5
   C. A Two-Part Method for Calculating Receptor Clustering Dynamics . 7
   D. Other Methods of Solving Aggregation Problems 10
   E. Synopsis 11

2. Branching Processes Applied to the Aggregation of f-Valent Particles .. 13
   A. The Galton-Watson Process 13
   B. Relationship Between the Galton-Watson Process and an
      Aggregation Process 15
   C. Probability Generating Functions 20
   D. Probability Generating Function for the Number of Offspring
      of an Individual in Generation r 20
   E. Weight Fraction Generating Function 22
   F. Lagrange Expansion 24
   G. Flory-Stockmayer Distribution 26

3. Multitype Branching Processes 28
   A. Antigen-Antibody Reactions 28
   B. Terminology and Definitions 29
   C. Weight Fractions 34

4. Aggregate Size Distribution on a Cell Surface 43
   A. General Considerations 43
   B. Bivalent Antigens 46
   C. Multivalent Antigens 50

TABLE OF CONTENTS (continued)

5. Gelation and Infinite-Sized Trees 59
    - A. General Considerations 59
    - B. Gelation in a One-Type Process 62
    - C. Gelation in a Two-Type Process (Antigen-Antibody Reactions) . . . 66
    - D. Infinite-Sized Aggregates on a Cell Surface 68
6. Post-Gel Relations 74
7. Conclusions and Extensions 84

Appendices 88
   - A. Proof of Theorem 2 88
   - B. Interactions of g-Valent Antibody with f-Valent Antigen 93
   - C. Generating Functions for Post-Gel Relations 100

List of Symbols 105

Bibliography 110

CHAPTER 1

INTRODUCTION

A. <u>RECEPTORS, LIGANDS, AND THE IMPORTANCE OF RECEPTOR CLUSTERING IN TRANSMEMBRANE SIGNALING</u>

Most cells have on their surfaces receptors that can specifically interact with molecules in the surrounding solution. Molecules that bind to receptors are collectively known as ligands. For example, the ligand may be a peptide hormone such as insulin and the receptor the "insulin receptor." Alternatively, the ligand might be a polysaccharide derived from the cell wall of a bacterium, and the receptor an immunoglobulin molecule on the surface of a lymphocyte, a type of white blood cell. If the ligand is a large molecule, it generally will have multiple sites, or functional groups, of which f can simultaneously interact with cellular receptors. Receptor molecules may also be multivalent and contain g, $g \geq 1$, identical ligand binding sites. For example, immunoglobulin molecules, which act as receptors on lymphocytes, basophils, and mast cells, contain two identical binding sites. When multivalent ligands interact with multivalent cell surface receptors, receptor-ligand aggregates form on the cell surface. The formation of such aggregates, as shown by a large number of biological examples, appears to be of primary importance in determining how cells respond to the presence of molecules in their environment. As we show later in this monograph, the size of receptor-ligand aggregates and their speed of formation depend on the ligand concentration in the surrounding solution (as well as other factors). Thus by sensing the size and/or rate of formation of receptor-ligand aggregates on its surface, a cell can roughly determine the concentration of ligand in its environment.

A number of recent papers and reviews have dealt with optimal ligand sensing mechanisms in the presence of noise (cf. Lauffenburger and DeLisi, 1983; Lauffenburger, 1982; DeLisi, Marchetti, and Del Grosso, 1982; DeLisi and Marchetti, 1983), and thus this topic will not be pursued here. Rather, our goal in this monograph is to solve a long-standing mathematical problem in the literature of receptor-ligand interactions: to compute, as a function of time, the size distribution of branched aggregates that result when f-valent ligands interact reversibly with g-valent receptors, and to determine the conditions under which such aggregates can grow infinitely large (i.e., form a gel).

As motivation for studying this problem, we note that the formation of ligand-receptor aggregates appears to be one of the first steps in transducing a signal from the membrane to the interior of many cells (Blum, 1985; King and Cuatrecasas, 1981; Loor, 1980; Schlessinger, 1979). Examples abound. Although the mechanisms by which insulin causes metabolic changes within a cell are not yet fully understood, one can show that the clustering of insulin receptors via bivalent anti-insulin antibody mimics the biological effects of insulin (Jacobs, Chang, and Cuatrecasas, 1978; Kahn, et al., 1978; Shechter, et al., 1979a). Similarly, the clustering of epidermal growth factor receptors or nerve growth factor receptors by anti-receptor antibody mimics the effect of the ligand (Shechter, et al., 1979b; Schreiber, et al., 1981). If instead of a bivalent anti-receptor antibody one utilizes a monovalent form of the antibody (i.e., an Fab fragment), then one finds the antibody has no biological effect (cf. Shecter, et al., 1979a).

One of the better-studied examples of the role of receptor-clustering in signal transduction occurs in the field of allergy. Basophils, which comprise approximately one percent of white blood cells in a normal individual, contain granules filled with histamine and other vasoactive substances. Mobile within the plane of the basophil cell membrane are Fc receptors which bind immunoglobulin E (IgE), a type of antibody molecule normally present in the blood plasma of all individuals. Normal individuals typically have 5,000-50,000 Fc receptors per basophil, whereas in allergic individuals one may find as many as 500,000 Fc receptors per basophil. The IgE molecules, once bound to Fc receptors, detect allergens such as pollen by binding them (Fig. 1.1). When the allergen is multivalent, more than one IgE may simultaneously bind the allergen, causing the normally mobile Fc receptors to cluster (Fig. 1.1). The clustering of Fc receptors provides a signal that leads to the degranulation of the basophil and the release of histamine (cf. Ishizaka, et al., 1984). Allergic individuals having elevated numbers of Fc receptors are more likely to form receptor clusters.

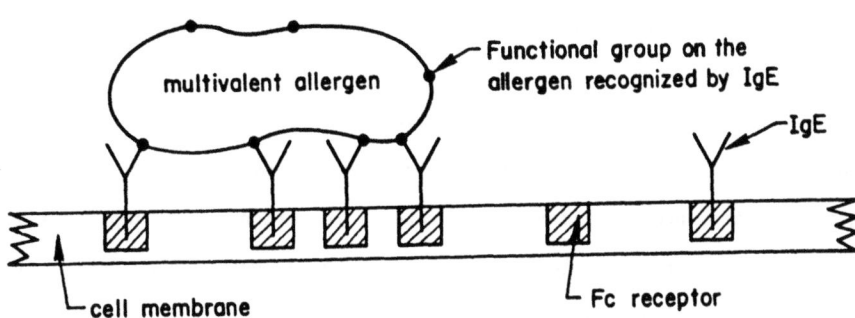

Figure 1.1

Fc receptors are mobile within the plane of the basophil membrane. These receptors bind IgE molecules which in turn bind allergen. If the allergen is multivalent, multiple IgE molecules can bind to it, thus clustering both the IgE molecules and the Fc receptors. This clustering event triggers degranulation and histamine release.

The hypothesis that the clustering of Fc receptors provides the signal for degranulation has been tested in many ways: 1) Monovalent allergens that bind to but can not cross-link surface IgE molecules do not cause degranulation; whereas multivalent allergens that can cross-link surface IgE molecules lead to degranulation and histamine release. 2) Bivalent antibodies raised against IgE (i.e., anti-IgE) cross-link surface IgE molecules and induce histamine release. 3) Anti-IgE molecules may be enzymatically cleaved so as to form two Fab fragments, each of which is monovalent and capable of binding to IgE. These Fab fragments do not induce histamine release. 4) Bivalent antibodies raised against the Fc receptor (anti-Fc receptor) cluster Fc receptors and stimulate basophil degranulation; Fab fragments of these antibodies generate no biological response. Thus antibodies which cross-link the Fc receptor either directly or via bound IgE can cause basophil degranulation in the absence of any allergen. This clearly is not the normal pathway of histamine release, but it does demonstrate the importance of receptor clustering in generating a transmembrane signal.

In each of the above three systems, simple binding of a ligand to a receptor is not sufficient to generate a cellular response; whereas the clustering of receptors, even by means not employing the ligand, can generate a response.

Unfortunately, it is not yet known which of the features of receptor-ligand aggregates -- size, number, lifetime, or rate of formation -- are important in determining a cell's response. In the case of basophils, MacGlashan, Schleimer, and Lichtenstein (1983) have shown, using various pharmocological inhibitors, that histamine release stimulated by trivalent antigens has different characteristics than release stimulated by bivalent antigens. Experiments and theory by Dintzis, Dintzis, and Vogelstein (1976), Vogelstein, Dintzis, and Dintzis (1982), Dintzis, Vogelstein, and Dintzis (1982), Dintzis, Middleton, and Dintzis (1983), Peacock and Barisas (1981a,b; 1983), and Barisas (1984) suggest that antibody secretion by B lymphocytes can be stimulated by receptor-ligand aggregates within a particular size-range and that the formation of cell surface aggregates outside this size range can lead to non-responsiveness and even tolerance of the cell. DeLisi (1981) suggests that the lifetime of a receptor cluster may be important in transmembrane signal amplification. Bell (1974) and Vogelstein, Dintzis, and Dintzis (1982) suggest that the rate of cluster formation is proportional to signal transduction. The results from different experimental groups are inconsistent and at times paradoxical (see Perelson, 1985b, for a review). Confusion is due at least in part to limitations of current technology which do not allow direct measurement of receptor cluster sizes. Only through the use of mathematical models (such as those presented in these notes) can one obtain information about the size and rate of formation of cell surface receptor clusters. By correlating theoretically predicted receptor states with experimentally observed biological responses, we hope to discover those features of receptor clusters that are important in determining cellular behavior.

B. <u>QUANTITATIVE THEORIES OF RECEPTOR CLUSTERING</u>

The problem of mathematically describing receptor clustering has been addressed previously. Perelson (1984) provides a tutorial review. Here we briefly describe the general characteristics of the existing models and refer the reader to the original literature for further details. Bell (1974) developed a simple two-step model for the binding of a multivalent ligand to a cell surface in which the ligand first reversibly attaches by a single functional group to a receptor and then by an irreversible reaction binds, via additional functional groups, to other receptors (Fig. 1.2). The second step in this process represents receptor cross-linking. Later, Bell (1975) generalized the model to include the competitive

---

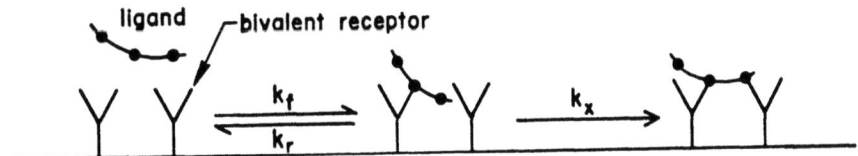

Figure 1.2

Bell's (1974) two-step model for the binding of a multivalent ligand to cell surface receptors. In the first step a ligand in solution reversibly binds to a single receptor. In the second step, a cell-bound ligand irreversibly binds to one or more additional receptors, creating cross-links between them.

---

effects of monovalent ligands that could bind the same cellular receptors as the multivalent ligand but could not cross-link them. DeLisi and Perelson (1976) and DeLisi and Thakur (1977) modified Bell's model but retained the basic two-step binding assumption. Dembo and Goldstein (1978), Wofsy, Goldstein, and Dembo (1978), Dembo, et al. (1978), Perelson (1979), DeLisi (1979), Perelson and DeLisi (1980), Perelson (1980), and Wofsy (1980) consider equilibrium and kinetic aspects of a more general model in which all binding and cross-linking steps are reversible, but restrict their attention to bivalent ligands. In these models, once a ligand reversibly binds to the surface with one of its two binding sites, it can, with its remaining binding site, reversibly cross-link receptors to form a linear chain or ring-shaped aggregate (Fig. 1.3). Vogelstein, Dintzis, and Dintzis (1982) develop a model for the reversible binding of a multivalent ligand to monovalent receptors. In this model up to f receptors can bind an f-valent ligand, and thus receptor clustering occurs. A related model has been developed by Dower et al. (1981a,b) for the binding of multivalent immune complexes to monovalent Fc receptors.

Work involving multivalent ligands with an effective valence $f > 2$ reversibly

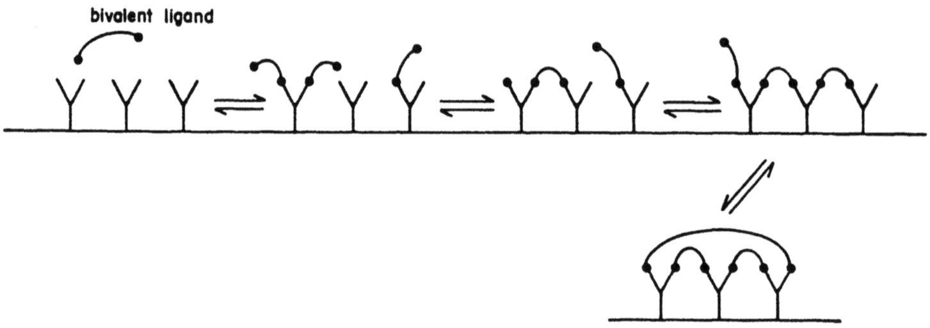

Figure 1.3

In the reaction of bivalent ligands with bivalent receptors, linear or ring-shaped receptor clusters can form.

---

binding to and cross-linking multivalent receptors is more scarce. These reactions are more difficult to study because they may lead to arbitrarily shaped networks. Gandolfi, Giovenco, and Strom (1978, 1979), DeLisi (1980), and Perelson (1981) develop a model (to be discussed in detail in Chapter 4) for the reversible binding and cross-linking of receptors by an f-valent ligand, $f \geq 2$, that involves determining the concentration of ligand molecules that have i sites bound, $i = 1, 2, \ldots, f$. Dower, Titus, and Segal (1984) extend this model to a population of cells that are heterogeneous with respect to receptor density. Using only the concentration of ligands that are multiply bound, various inferences have been made about cross-linking, but the entire distribution of cluster sizes has not been obtained (Perelson, 1981). The general cluster size distribution is computed for the first time in Chapter 4. A preliminary announcement of this result appears in Macken and Perelson (1982). Goldstein and Perelson (1984) discuss a special case -- the interaction of trivalent ligand with bivalent receptors -- and its applicability to histamine release experiments.

C. A TWO-PART METHOD FOR CALCULATING RECEPTOR CLUSTERING DYNAMICS

We use a two-part method to calculate the cluster size distribution. First we utilize a branching process technique previously developed by Gordon and coworkers (Gordon, 1962; Gordon and Scantlebury, 1964; Dobson and Gordon, 1964; Gordon and Malcolm, 1966) and Good (1963) for the purpose of computing aggregate size distributions in condensation polymerization reactions. For systems composed of g-valent receptors and f-valent ligands, the branching process method requires as *a priori* information the probability that at time t, k ($k = 0, 1, \ldots, g$) sites are bound on a receptor, as well as the probability that at time t, ℓ ($\ell = 0, 1, \ldots, f$) sites are bound on a ligand. Thus at any given time t, a set of $g + f$ independent site occupancy probabilities must be specified. As a prototypic example, which we shall use throughout the paper, we shall consider an antibody molecule as the receptor (i.e., $g = 2$) and an f-valent antigen as the ligand. With this in mind, we introduce the subscript A for antibody (receptor) and G for antigen (ligand). Consequently, in the branching process calculation one must specify $p_{Ak}(t)$ and $p_{G\ell}(t)$, $k = 0, 1, 2$ and $\ell = 0, 1, \ldots, f$, the respective probabilities of k antibody sites bound and ℓ antigen sites bound. The choice of $g = 2$ will not limit the generality of the method; it is chosen to allow us to make comparisons with previously studied problems and to simplify some of the algebraic expressions that arise. In Appendix B we derive results valid for all positive integer values of g.

The second part of our method involves selecting a model for the interaction of the ligand with cell surface receptors and using that model to calculate $p_{Ak}(t)$ and $p_{G\ell}(t)$. We illustrate this part of the method in Chapter 4 via the model of Gandolfi, Giovenco, and Strom (1978, 1979), DeLisi (1980), and Perelson (1981).

An enlightening way to introduce our method is to examine the aggregation of identical f-valent particles, a process which has been well studied in polymer chemistry (c.f. Stockmayer, 1943; Flory, 1953; Gordon, 1962; Good, 1963; Falk and Thomas, 1974; Donoghue and Gibbs, 1978, 1979; Donoghue, 1982, 1984; Ziff, 1980; Ziff and Stell, 1980; Cohen and Benedek, 1982). Particles having f identical sites interact, and the interaction of a free site on one particle (or cluster) with a free site on a second particle (or cluster) creates a bond which holds the particles (clusters) together. Bond formation may be reversible or irreversible; we consider the more general reversible case. To illustrate our method, let us assume that all free sites are equally reactive irrespective of the size of the cluster on which they are found. Thus a single forward rate constant, k_f, characterizes all association reactions, and a single reverse rate constant, k_r, characterizes all dissociation reactions. Furthermore, since all sites are equally reactive, the probability that any site is bound is independent of the state of all other sites.

To utilize the branching process method we must specify p_k, the probability that k sites are bound on a randomly chosen f-valent particle. Let p be the probability that a randomly chosen site is bound, i.e., the extent of reaction. Then, because sites act independently, p_k is given by the binomial formula, i.e.,

$$p_k = \binom{f}{k} p^k (1-p)^{f-k} \quad . \tag{1.1}$$

In order to study the dynamics of the aggregation process, we must specify how p_k, or equivalently p, changes with time. Assume we begin the process with a suspension of f-valent monomers at concentration C_0 and that aggregates collide at random and react according to the law of mass action. Because sites act independently, the concentration of free sites in the system at time t, S(t), is given by

$$\frac{dS}{dt} = -k_f S^2 + k_r(S_0 - S) \quad , \tag{1.2}$$

where $S(0) = S_0 = fC_0$ is the total concentration of sites in the system. This equation is easily solved (cf. DeLisi and Perelson, 1976). Once S(t) is known, it can be used to evaluate

$$p(t) = \frac{S_0 - S(t)}{S_0} \quad , \tag{1.3}$$

and hence, via Eq. (1.1), $p_k(t)$ can be found. In the next chapter we show, using branching processes, how the aggregate size distribution at any given time t is calculated, assuming $p_k(t)$ is known. Further, for tutorial purposes, we shall show in Chapter 3 how this method can be extended to study the aggregation of two different types of molecules (such as antibodies and antigens) in solution. Although these simple examples are not our endpoint, they have applicability in many practical situations such as the typing of blood.

More complicated models are necessary to study the binding of ligands to cells. We cannot assume that all ligand sites are equally reactive irrespective of the ligand's state of aggregation because the binding of a ligand to the cell surface is quite different in character and has different rate constants from the cross-linking reactions that a cell-bound ligand undergoes. However, the branching process part of the calculation will be totally independent of the ligand binding dynamics once a few basic assumptions are made, e.g., whether or not intramolecular reactions are allowed.

D. OTHER METHODS OF SOLVING AGGREGATION PROBLEMS

Because we believe that the methods we develop in these notes will prove to be useful in analyzing a large variety of aggregation problems, we discuss the mathematical basis of our approach in detail and show how the method is used by analyzing some classical problems approached earlier by different methods. Aggregation processes have been studied by a variety of mathematical techniques. As discussed in the Preface, Smoluchowski (1917) introduced an infinite system of ordinary differential equations to describe the irreversible condensation of colloid droplets. Differential equation models are still used (cf. Samsel and Perelson, 1982, 1984; Seinfeld, 1980) and generalizations for reversible processes have been proposed (Perelson and DeLisi, 1975, Samsel and Perelson, 1984, Dongen and Ernst, 1983, 1984). Continuous approximations to the infinite system of ordinary differential equations have been used in a number of fields (cf. Liu and Amundson, 1962; Zeman and Amundson, 1963; Aris and Gavalas, 1965; Seinfeld, 1980; Falkovitz and Segel, 1982). A stochastic method based upon equal reactivity of all sites and the absence of intramolecular reactions was pioneered by Flory (1936, 1941a,b,c). Stockmayer (1943, 1944, 1952) used a statistical mechanical technique to compute most probable aggregate size distributions. Goldberg (1952, 1953) applied Stockmayer's method to the study of antigen-antibody aggregation in solution. As mentioned previously, Gordon and coworkers (Gordon, 1962; Gordon and Scantlebury, 1964; Dobson and Gordon, 1964; Gordon and Malcolm, 1966) used branching processes to derive molecular weight distributions of polymers. Good (1963) made explicit some of the mathematical underpinnings of Gordon's work and rederived the Stockmayer (1943) result for the weight fraction distribution of aggregates formed by the random condensation of f-valent particles. Lowry (1970), Whittle (1972), and Kelly (1979) showed how Markov chains could be used to analyze this class of problems. Recently, techniques involving percolation theory, fractals, the theory of critical phenomena, and extensive computer simulation have been applied to the study of aggregation phenomena. Family and Landau (1984) provide an introduction to these new approaches.

Much of this previous work relies on the two assumptions originally made by Flory (1953) and Stockmayer (1943). First, all sites of a given type (e.g., antigen sites) are equally reactive irrespective of the size of the aggregate on which they are located. This implies that a single forward and a single reverse rate constant can be used to describe all the reactions involving a particular type of site. Diffusion-limited reactions in which the rate of reaction of a site depends on the diffusion rate of the aggregate to which it is attached would be excluded (cf. Cohen and Benedek, 1982), as well as reactions in which steric hindrance might block sites in the interior of an aggregate.

Second, the classical results rely on the assumption that no intramolecular reactions occur. This implies that two free sites on the same aggregate never interact, and hence there cannot be any loops or cycles in the aggregate. Topologically the aggregate will resemble a tree. The branching process approach developed by Gordon (1962) was important because it allowed one to relax the equal reactivity assumption. Further, Gordon and Scantlebury (1966) showed how one could use branching processes to obtain approximate information about aggregation processes even when cyclization reactions are allowed. A general kinetic theory of branched aggregations that includes cyclization reactions still remains to be developed.

E. SYNOPSIS

In the work that follows, we restrict our attention to aggregates that can be represented graphically as a tree. In our theory, "trees" become schematic diagrams for branching processes, and the mathematical machinery of generating functions can be used to answer questions concerning tree composition. Readers familiar with other methods for analyzing aggregation processes will quickly appreciate the ease with which difficult combinatorial problems, which arise in the Flory and Stockmayer methods, are solved through the generating function approach.

In Chapter 2 a brief history of branching processes is given. Then probability generating function techniques are introduced and applied to the aggregation of f-valent particles. Stockmayer's (1943) result for weight fraction of n-mer is obtained in a surprisingly simple fashion. Chapter 3 extends the use of generating functions to the aggregation of two types of particles. Goldberg's (1952) results are obtained for the weight-fraction of aggregates in solution composed of i bivalent antibodies and j f-valent antigens. Goldberg's (1953) result for the size distribution of aggregates composed of f-valent antigen and g-valent antibody is derived in Appendix B. Although these results are not new, they are included for pedagogical reasons because they illustrate, in a simple fashion, the techniques needed to analyze ligand-receptor interactions.

In Chapter 4 we use the methods developed in Chapters 2 and 3 to derive the cluster size distribution for receptor-ligand aggregates on a cell surface. This result is new and extends the previous work of DeLisi (1980), Perelson (1981), and Goldstein and Perelson (1984).

Chapters 5 and 6 deal with the formation of infinite-sized aggregates or gels. In Chapter 5 criteria for gelation are derived for systems of f-valent particles, for antigen-antibody systems, and for receptor-ligand systems. In Chapter 6 we derive the weight fraction distribution of the finite-sized aggregates belonging to the sol in a system containing a gel phase. When applied to cell surface aggregation phenomena, these results provide rigorous means for predicting the onset of gelation and the aggregate size distribution in the post-gel regime. Chapter 7 summarizes our results and discusses possible extensions and other applications of the branching process technique in aggregation phenomena.

CHAPTER 2

BRANCHING PROCESSES APPLIED TO THE AGGREGATION OF f-VALENT PARTICLES

A. THE GALTON-WATSON PROCESS

<u>2.1</u> The earliest formulation of a branching process is usually attributed to Sir Francis Galton and the Reverend H. W. Watson who, being concerned with the extinction of prominent English families, developed the generating function methodology to critically examine family trees and determine conditions for the eventual loss of a surname (cf. Harris, 1963). As a point of historical interest, it has recently been learned (Heyde and Seneta, 1972) that Galton and Watson were anticipated by almost 30 years by Bienaymé, an obscure French mathematician (see Kendall, 1975).

The Galton-Watson process, as it has become known, is generally used to study objects that, at the end of their "lifetime," can generate additional objects of the same or related types in a stochastic manner. The objects might be whole organisms or genes or even neutrons in a chain reaction. In the Galton-Watson process there is an initial population of Z_0 individuals, called the zeroth generation, each of whom reproduces independently of the others with probability p_k of bearing k children, where

$$p_k \geq 0 \ , \quad k = 0, 1, 2, \ldots \quad \text{and} \quad \sum_{k=0}^{\infty} p_k = 1 \ . \tag{2.1}$$

The "offspring" of all the individuals in the zeroth generation constitutes the first generation, whose size we denote by Z_1. Each individual in the first generation, independently of all other individuals in the population, bears progeny in accordance with the probability distribution given by Eq. (2.1). This collection of offspring constitutes the second generation, with size Z_2. The Galton-Watson process continues in a like manner so that the rth generation, of size Z_r, is composed of descendants of the (r-1)st generation. Further, each member of the rth generation independently produces k offspring with probability p_k, k = 0, 1, 2, ... As shown in Fig. 2.1, the Galton-Watson process for $Z_0 = 1$ can be represented graphically by a rooted tree or "family tree," the root being an individual in generation zero and Z_r being the number of nodes at distance r from the root.

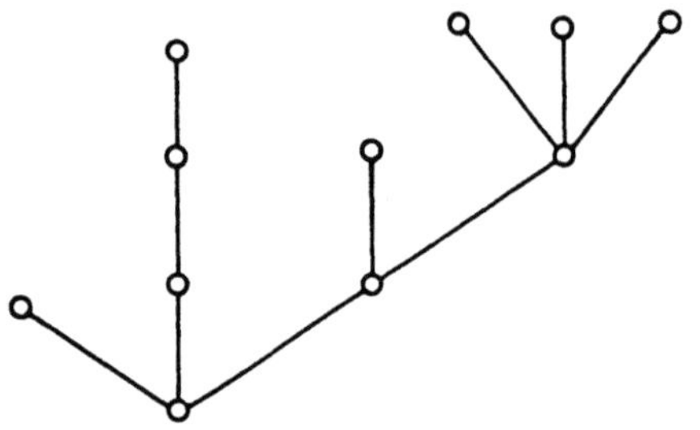

generation = r	Z_r	Y_r
4	0	11
3	4	11
2	3	7
1	3	4
0	1	1

Figure 2.1

A Galton-Watson process represented as a rooted tree, the root being the node in generation 0. The number of individuals (nodes) in generation r, Z_r, is indicated as well as the total number of individuals in the tree through the rth generation, Y_r. Here the tree (branching process) goes extinct by generation 4 since parents in generation 3 produce no offspring. Note Y_r attains its limiting value, Y, when the tree goes extinct.

2.2 Let us assume $Z_0 = 1$, i.e., the zeroth generation consists of a single individual. Then if Y_r denotes the total size of the family descended from that individual by the rth generation (i.e., the individual in the zeroth generation plus the total number of its descendants through the rth generation),

$$Y_r = \sum_{i=0}^{r} Z_i \quad . \tag{2.2}$$

If for some n, $Z_n = 0$, then the population has gone <u>extinct</u>, and obviously $Z_k = 0$ for all $k > n$. Further, under these conditions the total number of individuals in a family tree,

$$Y = \lim_{r \to \infty} Y_r \quad , \tag{2.3}$$

is finite. If the family does not go extinct, then Y is an infinite-valued random variable. Y is commonly called the <u>total progeny of the branching process</u> (Jagers, 1975).

Let q be the probability of extinction, i.e.,

$$q = P(Z_n = 0 \text{ for some } n) \quad . \tag{2.4}$$

Mullikin (1968) showed that this is equivalent to defining q as $P(Y < \infty)$.

B. RELATIONSHIP BETWEEN THE GALTON-WATSON PROCESS AND AN AGGREGATION PROCESS

<u>2.3</u> In the two sections following, we will show how to construct a modified version of the Galton-Watson process and then use it to determine the size and weight fraction distributions of aggregates composed of f-valent particles. We restrict our attention to aggregates that contain no cyclic structures and hence have the topological form of a tree.

<u>2.4</u> In constructing an analogy between Galton-Watson family trees and aggregates having a treelike form, there are several conventions that we must adopt. First, we equate p_k, the probability of having k offspring, with the probability of having one particle bound to k others. Second, the particle valence in the aggregation process must be accounted for in the Galton-Watson process by imposing a restriction on the maximum possible number of offspring contributed by a single parent to the next generation. Thus a parent in the zeroth generation can have at

most f offspring, whereas a parent in later generations can have at most f-1 offspring because one particle site is used to attach the particle to the aggregate (see Fig. 2.2). Third, although the Galton-Watson terminology of "parent," "offspring," and "generation" are applied to the aggregation process, no generations exist as such since we are studying aggregates at some fixed time t. (As we mentioned in the introduction, we use a separate kinetic model to relate our analysis at time t to the aggregate size distribution at other times.) In order to assign the nodes of a treelike aggregate to various "generations," one can arbitrarily pick one node (particle) and regard it as the initial "parent" in the zeroth generation. In graph theory this distinguished node is known as the <u>root</u> of the tree (Harary, 1969). Once the root is chosen, the connectedness of the tree assigns all other nodes to the appropriate generation, as shown in Fig. 2.3. Because there is no natural choice for the root, any of the particles in an aggregate can serve as the root. Thus any of n possible rooted trees can represent an n-mer. As we discuss below, this degeneracy is taken into account when we compute the relative abundance of n-mers in a population of aggregates.

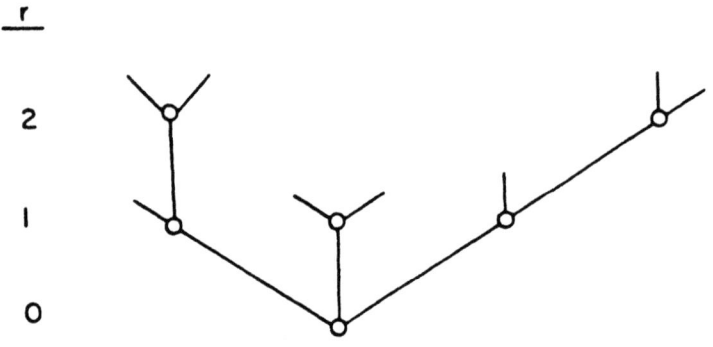

Figure 2.2

A typical family tree representing the aggregation of f-valent particles. Here f = 3. Notice particles in generation r = 0 can have at most f offspring, whereas in all later generations a particle can have at most f-1 offspring.

To summarize, the basis of the analogy between Galton-Watson processes and aggregation processes is that an n-mer is represented by a rooted tree containing n nodes, with the degree of the root being at most f and the degree of all other nodes being at most f-1.

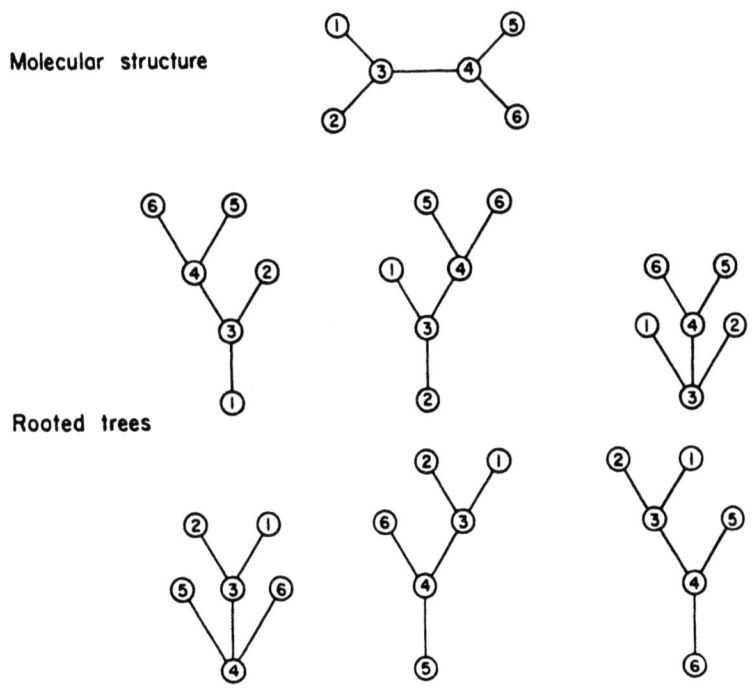

Figure 2.3

An aggregate containing six labeled particles and the six distinct rooted trees that correspond to it.

2.5 We elaborate here on a point that is critical to understanding the use of branching processes in calculating weight fraction distributions. In nature, the relative abundance of an aggregate at equilibrium is proportional to its energy, via a Boltzmann factor, and a statistical mechanical degeneracy factor that measures the number of equivalent aggregates that can be built out of n monomers. In kinetic theories, similar degeneracy factors multiply the forward and reverse rate constants and account for the number of different ways each reaction would occur. Gordon and coworkers (Gordon, 1970; Gordon and Judd, 1971; Gordon and Temple, 1970, 1973, 1976; Gordon with Leonis, 1975) establish with some rigor that the branching process method correctly counts the number of equivalent aggregates, assigns them the correct energy, and hence predicts the weight fraction distribution as found in nature. Rather than invoking statistical mechanical principles involving partition functions, distinguishability of particles, symmetry groups, etc. (which are outside the scope of this monograph), we give here a heuristic argument and later some explicit examples to demonstrate that the method gives correct results. The interested reader may refer to the aforementioned papers by Gordon and coworkers.

In the branching process method, the energy of an aggregate is taken to be the sum of the energies of the various bonds holding the aggregate together. The Boltzmann factor is therefore included in the offspring probabilities that represent the probabilities of forming chemical bonds. To obtain the degeneracy factor, we note that the branching process method will generate all possible rooted trees consistent with the valence of the monomers. In this collection of trees, many will have exactly n nodes. The statistical mechanical degeneracy factor corresponds to the size of this collection divided by n. The factor of n arises because there are n times as many rooted trees as there are distinguishable n-mer structures.

We now make the crucial observation that allows us to calculate a weight fraction distribution using branching processes. If at some time in an aggregation process the extent of reaction and probabilities p_k are such that there exists N_n n-mers in the system, then the corresponding branching process with offspring probabilities p_k will generate nN_n rooted trees of size n.

$$P(\text{rooted tree of size n}) = \frac{nN_n}{\sum_{k=1}^{\infty} kN_k} \quad . \tag{2.5}$$

If the weight of an f-valent particle (monomer) is taken to be the unit of weight, then each n-mer will have weight n, and w_n, the weight fraction of n-mer, is given by

$$w_n \equiv \frac{nN_n}{\sum_{k=1}^{\infty} kN_k} \quad . \tag{2.6}$$

Comparing this with Eq. (2.5) we find

$$w_n = P(\text{rooted tree of size n}) \quad . \tag{2.7}$$

Recalling that Y is the total number of individuals in a family tree, Eq. (2.7) can be rewritten as

$$w_n = P(Y = n) \quad . \tag{2.8}$$

To this point, we have assumed the collection of aggregates contains only finite-sized members. If an infinite-sized aggregate is present -- as occurs in gelation -- it will be represented by a set of infinite-sized rooted trees, i.e., family trees which have not gone extinct. Since $\Sigma k N_k$ does not include contributions "at infinity," $nN_n/\Sigma kN_k$ defines probabilities and weight fractions conditional upon being in the finite portion of the system. Branching processes can then be used to study the composition of the sol fraction of the system. We defer a discussion of gelation until Chapter 5 and continue to assume all aggregates are finite.

In summary, we see that the problem of computing the weight fraction of n-mers reduces in the branching processes approach to calculating the probability distribution $P(Y = n)$ for rooted finite trees. The most efficient way to calculate such a probability distribution was discovered by the Reverend Watson in 1874 and involves the use of probability generating functions.

C. PROBABILITY GENERATING FUNCTIONS

2.6 By definition, an integer-valued non-negative random variable X has a <u>probability generating function</u>, $F(\theta)$, given by

$$F(\theta) = \sum_{k=0}^{\infty} p_k \theta^k \quad , \tag{2.9}$$

where

$$p_k = P(X = k) \tag{2.10}$$

and θ is a dummy complex variable. For $p_k \geq 0$ and $\sum_{k=0}^{\infty} p_k = 1$, $F(\theta)$ is defined at least for $|\theta| \leq 1$ and is infinitely differentiable for $|\theta| < 1$. If X is the number of offspring of an individual in a Galton-Watson process, then the generating function $F(\theta)$ is a compact representation of the offspring distribution $P(X = k) = p_k$. Given $F(\theta)$, one can recover the probability distribution, i.e.,

$$p_k = \frac{1}{k!} \left. \frac{d^k F(\theta)}{d\theta^k} \right|_{\theta=0} \quad , \quad k = 1, 2, \ldots \quad . \tag{2.11}$$

Further, the derivative of the generating function reduces to the expected value of X when evaluated at $\theta = 1$, i.e.,

$$E(X) = F'(1) \quad . \tag{2.12}$$

D. PROBABILITY GENERATING FUNCTION FOR THE NUMBER OF OFFSPRING OF AN INDIVIDUAL IN GENERATION r

2.7 Recall that in treating the aggregation of f-valent particles as a branching process, a particle in the zeroth generation can have at most f offspring; whereas a particle in later generations can have at most f-1 offspring. Consequently, the generating function for the number of offspring of an individual in generation r, $F_r(\theta)$, must differ for the cases $r = 0$ and $r \geq 1$. With this notation,

$$F_0(\theta) = \sum_{k=0}^{f} p_k \theta^k \tag{2.13}$$

where p_k is the probability of an individual in generation 0 having k offspring. If all sites are equally reactive, then p_k is given by Eq. (1.1), and

$$F_0(\theta) = (1-p + p\theta)^f \quad. \tag{2.14}$$

Analogously,

$$F_1(\theta) = \sum_{k=0}^{f-1} \tilde{p}_k \theta^k \quad, \tag{2.15}$$

where $\tilde{p}_k$ is the probability that an individual in the first generation has k offspring. If we assume a particle in the first generation behaves chemically in a manner identical to a particle in the zeroth generation, with the exception that it is already bound at one site to the particle in the zeroth generation, then

$$\tilde{p}_k = \frac{(k+1)p_{k+1}}{\sum_{n=0}^{f-1} (n+1)p_{n+1}} \quad, \quad k = 0, 1, \ldots, f-1 \quad. \tag{2.16}$$

Equation (2.16) arises from the fact that a particle in the first generation with k offspring has a total of k+1 bound sites. This occurs with probability p_{k+1}. Further, any of these k+1 sites may connect the particle to the zeroth generation. The denominator in Eq. (2.16) correctly normalizes $\tilde{p}_k$. As a consequence of Eq. (2.16)

$$F_1(\theta) = \frac{F_0'(\theta)}{F_0'(1)} \quad, \tag{2.17}$$

where the prime denotes differentiation. Observe that by Eq. (2.12), $F_0'(1)$ is just the mean number of offspring of an individual in the zeroth generation.

<u>2.8</u> <u>Remark</u> Gordon and Malcolm (1966) call Eq. (2.17) a <u>universal consistency relation</u>. This relation was also noted by Whittle (1965a) and Gordon and Scantlebury (1964).

<u>2.9</u> If all sites are equally reactive, then as a consequence of Eq. (2.17),

$$F_1(\theta) = (1-p + p\theta)^{f-1} \tag{2.18}$$

as one would expect.

<u>2.10</u> The simplest assumption to make about the generating functions $F_2(\theta)$, $F_3(\theta)$, ..., is that

$$F_r(\theta) = F_1(\theta) \quad , \quad r = 2, 3, \ldots \quad . \tag{2.19}$$

This is equivalent to assuming that the probability that an individual in generation r has k offspring is equal for all generations except the zeroth. In the following, we assume that Eq. (2.19) holds.

E. <u>WEIGHT FRACTION GENERATING FUNCTION</u>

<u>2.11</u> As formulated in Section B of this chapter, the major problem in applying branching processes to the computation of the number or weight fraction distribution of aggregates is to determine the probability distribution $P(Y = n)$. To do so, we introduce $W_r(\theta)$, the generating function of Y_r, the total number of progeny in a tree up to and including the rth generation, i.e.,

$$W_r(\theta) = \sum_{k=0}^{\infty} P(Y_r = k)\theta^k \quad . \tag{2.20}$$

In particular,

$$W_1(\theta) = \sum_{k=0}^{\infty} P(Y_1 = k)\theta^k \quad .$$

By Eq. (2.2), $Y_1 = 1 + Z_1$ (recall $Z_0 = 1$), and thus

$$W_1(\theta) = \theta \sum_{k=1}^{\infty} P(Z_1 = k - 1)\theta^{k-1}$$

$$= \theta F_0(\theta) \quad . \tag{2.21}$$

Rather than computing $W_2(\theta)$, $W_3(\theta)$, etc., we simply quote Good (1949) who has shown that $W_r(\theta)$ is given by the following composition:

$$W_r(\theta) = \theta F_0(\theta F_1(\theta F_2(\ldots \theta F_{r-1}(\theta)))) \quad . \tag{2.22}$$

Under the assumption of equal generating functions for all generations after the zeroth, cf. Eq. (2.19),

$$W_r(\theta) = \theta F_0(u_{r-1}(\theta)) \quad , \quad r = 2, 3, \ldots \quad , \tag{2.23}$$

where $u_{r-1}(\theta)$ is θF_1 composed with itself $r-1$ times.

Recall that an n-mer is represented by a set of n family trees all of which have gone extinct before the nth generation. To ensure that we include all finite family trees in our counting, we examine $W(\theta)$, the limit of $W_r(\theta)$ as $r \to \infty$. $W(\theta)$ is the weight fraction generating function from which the aggregate size distribution can be obtained. We now study this generating function in detail. From Eq. (2.23),

$$W(\theta) = \lim_{r \to \infty} W_r(\theta) = \theta F_0(u(\theta)) \quad , \tag{2.24}$$

where $u(\theta)$ is the limit of $u_r(\theta)$ as $r \to \infty$. One can show (Feller, 1968; Good, 1949, 1960) that $u_r(\theta)$ is a decreasing function of r for each θ ($0 < \theta < 1$), and hence the limit $u(\theta)$ exists. Since $u_r(\theta) = \theta F_1(u_{r-1}(\theta))$, the limit $u(\theta)$ satisfies

$$u(\theta) = \theta F_1(u(\theta)) \quad , \quad 0 < \theta < 1 \quad . \tag{2.25}$$

For any given value of θ between 0 and 1, u is therefore a root of

$$u = \theta F_1(u) \quad . \tag{2.26}$$

Although we shall not do so here, one can show that the limit of u_r must be the smallest positive root of Eq. (2.26).

Combining Eqs. (2.24), (2.20), and (2.8), we find the fundamental relationship

$$W(\theta) = \sum_{n=0}^{\infty} w_n \theta^n = \theta F_0(u(\theta)) \quad , \tag{2.27}$$

where w_n is the weight fraction of n-mer. Following Good (1963), we introduce the notation $C(\theta^n)\{W(\theta)\}$ for the coefficient of θ^n in the expansion of $W(\theta)$. Thus

$$w_n = C(\theta^n)\{W(\theta)\} \quad . \tag{2.28}$$

F. **LAGRANGE EXPANSION**

2.12 Writing $F_0(u(\theta))$ as an explicit function of θ and then extracting the coefficient of θ^n is complicated because first $u(\theta)$ must be found by solving Eq. (2.26). Rather than approach the problem this way, we can use Lagrange's expansion of an inverse function (cf. Jeffreys and Jeffreys, 1972; Bromwich, 1947; Whittaker and Watson, 1935; Goursat, 1904) to construct from $\theta = u/F_1(u)$, u as a power series in θ. Similarly by Lagrange expansion, any function of u, and in particular $F_0(u)$, can be written as a power series in θ (cf. Jeffreys and Jeffreys, 1972; p. 383). The following theorem from Good (1963) determines the coefficient of θ^n in this power series. The notation $F^n(u)$ is used to denote $F(u)$ raised to the nth power; $F'(u)$ denotes the derivative of $F(u)$.

Theorem 1 For $F_0(u)$ and $F_1(u)$ analytic near the origin, $F_1(0) \neq 0$ and $\theta = u/F_1(u)$,

$$C(\theta^n)\{F_0(u(\theta))\} = n^{-1} C(u^{n-1})\{F_0'(u) F_1^n(u)\} \quad . \tag{2.29}$$

2.13 **Remark** The restriction $F_1(0) \neq 0$ has the interpretation that the probability of zero offspring for a parent in the first or subsequent generations can not be zero, i.e., $\tilde{p}_0 \neq 0$. If this probability were zero, then a branching process that reached the first generation would never end and hence would generate an infinite tree. Because $F_0(u)$ and $F_1(u)$ are defined as probability generating functions, they necessarily are analytic near the origin; in fact, they are analytic for all $|u| < 1$.

2.14 **Proof** Theorem 1 will play a central role in all that follows. Therefore, we give here a simple proof, based on Bromwich (1947).

To determine the coefficient of θ^n in the function $F_0(u(\theta))$, we first express θ as a power series in u. Such a power series exists since $\theta = u/F_1(u)$ and $F_1(u)$ is analytic and thus expressible as a power series in u. Observe that $F_1(0) \neq 0$, and therefore $\theta = 0$ at $u = 0$. Hence a power series expansion of $u/F_1(u)$ has no constant term. We write

$$\theta = \sum_{r=1}^{\infty} a_r u^r \quad . \tag{2.30}$$

Equation (2.30) may be solved for u as a power series in θ, and thus any function of u, say $F_0(u)$, can be expressed as a power series in θ. Let

$$F_0(u(\theta)) = \sum_{r=0}^{\infty} b_r \theta^r \quad . \tag{2.31}$$

The proof now amounts to showing

$$nb_n = C(u^{n-1})\{F_0'(u)F_1^n(u)\} \quad .$$

Since $\theta = u/F_1(u)$, this is equivalent to showing

$$nb_n = C(u^{n-1})\{F_0'(u)u^n/\theta^n\}$$

$$= C(u^{-1})\{F_0'(u)/\theta^n\} \quad .$$

From Eq. (2.31) we obtain

$$\frac{F_0'(u)}{\theta^n} = \sum_{r=1}^{\infty} rb_r \theta^{r-n-1} \frac{d\theta}{du} \quad . \qquad (2.32)$$

If $r \neq n$, then

$$\theta^{r-n-1} \frac{d\theta}{du} = \frac{1}{r-n} \frac{d}{du} (\theta^{r-n})$$

$$= \frac{1}{r-n} \frac{d}{du} [u^{r-n}(A_0 + A_1 u + \ldots)] \quad , \qquad (2.33)$$

where the last equality results from Eq. (2.30). Because Eq. (2.33) involves the derivative of a power series which does not contain any terms proportional to $\ln(u)$, $\theta^{r-n-1} d\theta/du$ will not contain a term in u^{-1}. On the other hand, if $r = n$, then

$$\frac{1}{\theta} \frac{d\theta}{du} = \frac{a_1 + 2a_2 u + 3a_3 u^2 + \ldots}{a_1 u + a_2 u^2 + a_3 u^3 + \ldots}$$

$$= \frac{1}{u} + B_1 + 2B_2 u + \ldots \quad ,$$

where we use Eq. (2.30) to obtain the first equality. Hence, from Eq. (2.32)

$$C(u^{-1})\{F_0'(u)/\theta^n\} = nb_n \quad ,$$

thus proving the Theorem.

G. **FLORY-STOCKMAYER DISTRIBUTION**

<u>2.15</u> <u>Example</u> As an example of the use of Theorem 1, we derive the Flory (1936, 1941b) - Stockmayer (1943) formula for the weight fraction of n-mers composed of f-valent particles with (assumed) equally reactive sites. Let p be the probability that a site is bound. Then by Eqs. (2.14), (2.18) and (2.27),

$$F_0(u) = (1-p + pu)^f \quad , \quad F_1(u) = (1-p + pu)^{f-1} \quad ,$$

and

$$W(\theta) = \theta F_0(u(\theta)) = \theta(1-p + pu(\theta))^f \quad .$$

Now,

$$C(\theta^n)\{W(\theta)\} = C(\theta^{n-1})\{F_0(u(\theta))\} \quad ,$$

which by Theorem 1

$$= (n-1)^{-1} C(u^{n-2})\{fp(1-p + pu)^{f-1}(1-p + pu)^{(f-1)(n-1)}\}$$

$$= fp(n-1)^{-1} C(u^{n-2})\{(1-p + pu)^{(f-1)n}\} \quad .$$

Binomial expansion of the term in braces gives the Flory-Stockmayer result,

$$w_n = \frac{f(fn-n)!}{(n-1)!(fn-2n+2)!} p^{n-1}(1-p)^{fn-2n+2} \quad . \tag{2.34}$$

CHAPTER 3

MULTITYPE BRANCHING PROCESSES

A. ANTIGEN-ANTIBODY REACTIONS

<u>3.1</u> The concepts and results of Chapter 2 are readily extended to aggregates of more than one particle type. Our interest lies in the aggregation behavior of f-valent antigen and bivalent antibody. Processes involving more than two types, [for example, f-valent antigen in solution with either a mixture of bivalent antibody and Fab fragments (univalent antibody), or a mixture of different affinity antibodies] are treated similarly, but the labor involved in carrying out the mathematical details increases dramatically. Thus, our approach will be to present the theory in its two-type form with references to the appropriate general version. Although in the main body of the text we restrict our development to bivalent antibody, in Appendix B we provide the appropriate generalizations for g-valent antibody.

Antigen-antibody reactions are most simply studied as they occur in solution, especially when one can assume that all sites are equally reactive. In other words we assume, for example, that the presence of one bond to an antigen has no effect on the reactivity of the remaining unbound sites of that antigen. (In chemical terminology we say there are no substitution effects such as steric hindrance.) Under these conditions, probability models for the number of sites bound particularize to functional forms which are mathematically more tractable than probability models for aggregation on a cell surface. The latter situation requires the full generality of probability generating functions, as we shall demonstrate in Chapter 4.

In this section we parallel the development of Chapter 2. The principle results we obtain are the weight fractions for a two-type process with general branching probabilities, from which we derive Goldberg's (1952) result on the size distribution of antigen-bivalent antibody aggregates in solution. Goldberg's (1953) generalization for the size distribution of aggregates formed between multivalent antibody and multivalent antigen is derived in Appendix B.

B. TERMINOLOGY AND DEFINITIONS

<u>3.2</u> We first show how antigen-antibody aggregates can be described by family trees containing more than one particle type. In antigen-antibody reactions, each type of particle may be a "parent," but a peculiarity of this process is that only "offspring" of the type opposite to the "parent" may be born.

Graphically, an aggregate may be represented as in Fig. 3.1 where the nodes denote antigens and the lines are antibodies. In contrast to Fig. 2.3, lines must now be counted as part of the aggregate and a line (antibody) may be the last generation.

Assuming that within a given generation parents of either type act independently of each other and of their past, the aggregation process may be modeled as a multitype branching process. Again, generating functions will prove to be effective tools for deriving weight fractions.

The root (zeroth generation) of a tree may be an individual of either type. We will assume forthwith that fertility of either type of "parent" is independent of generation after the zeroth. [The same assumption was made in Chapter 2, Eq. (2.19).] In terms of our antigen-antibody aggregates, the assumption is that a free antigen (antibody) site in the interior of the cluster is as reactive as a free antigen (antibody) site at the extremity of the cluster. If there are differences in reactivity between internal and external sites, then different generating functions will have to be assigned to the different generations, rendering the theory all but intractable.

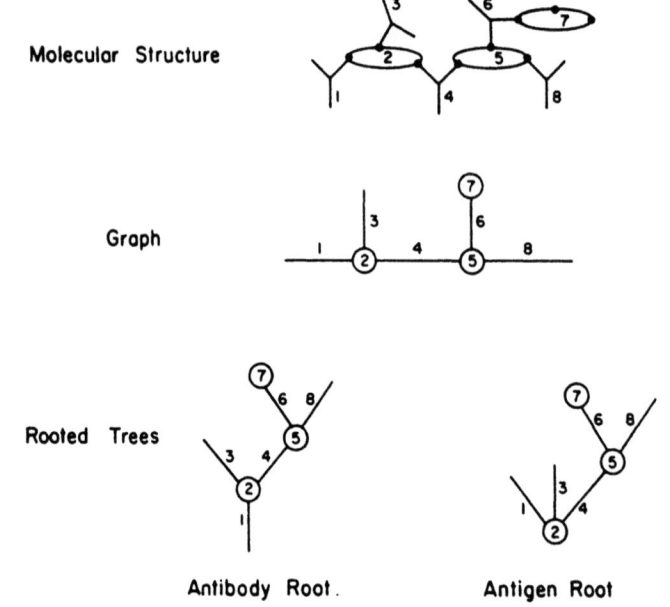

Figure 3.1

Correspondence between the molecular structure of an antigen-antibody aggregate and a collection of rooted trees. Antibodies are Y-shaped molecules with two identical antigen binding sites, one located at the end of each arm of the Y. Antigens are depicted here as globular molecules with three sites (dark dots) to which antibodies can attach. The molecular structure is converted into a graph by replacing each antigen with a node and each antibody with a branch. The graph is then converted into a collection of rooted trees by choosing each antibody and each antigen in turn as a root. Two representative rooted trees are shown.

3.3 We now define the bivariate probability generating function that characterizes this branching process. If "A" denotes "antibody" and "G" denotes "antigen," let

$p_{\ell k}^{(i)}$ = P(root of type i produces ℓ offspring of type A and k offspring of type G) , i = A, G . (3.1)

Then the vectorial generating function for offspring from a root of either type will be defined as

$$\underset{\sim}{F}_0(\underset{\sim}{\theta}) = (F_{A0}(\underset{\sim}{\theta}), F_{G0}(\underset{\sim}{\theta})) \quad , \tag{3.2}$$

where

$$\underset{\sim}{\theta} = (\theta_A, \theta_G) \quad , \tag{3.3}$$

θ_A and θ_G being dummy complex variables, $|\theta_A| < 1$, $|\theta_G| < 1$ and

$$F_{i0}(\underset{\sim}{\theta}) = \sum_{k=0}^{\infty} \sum_{\ell=0}^{\infty} p_{\ell k}^{(i)} \theta_A^\ell \theta_G^k \quad , \quad i = A, G \quad . \tag{3.4}$$

We shall simplify our notation by observing that if the root is of type A then all offspring must be type G, i.e.,

$$p_{\ell k}^{(A)} = 0 \quad , \quad \ell \geq 1 \quad , \quad k \geq 0 \tag{3.5a}$$

and conversely, if G is the root, then all offspring must be type A, i.e.,

$$p_{\ell k}^{(G)} = 0 \quad , \quad \ell \geq 0 \quad , \quad k \geq 1 \quad . \tag{3.5b}$$

Thus, we can replace Eq. (3.1) by

$$p_{Ak} = P(\text{a root of type A produces k offspring of type G}) \quad ,$$
$$k = 0, 1, 2 \tag{3.6a}$$

and

$$p_{G\ell} = P(\text{a root of type G produces } \ell \text{ offspring of type A}) \quad ,$$
$$\ell = 0, 1, \ldots, f \quad . \tag{3.6b}$$

The generating functions in Eq. (3.4) now reduce to

$$F_{A0}(\underset{\sim}{\theta}) = \sum_{k=0}^{2} p_{Ak} \theta_G^k \tag{3.7a}$$

and

$$F_{G0}(\underset{\sim}{\theta}) = \sum_{\ell=0}^{f} p_{G\ell}\theta_A^\ell \quad . \tag{3.7b}$$

The "0" subscript on F denotes the zeroth generation (root).

For generations one, two, three, ..., we define

$$F_{A1}(\underset{\sim}{\theta}) = \sum_{k=0}^{1} \tilde{p}_{Ak}\theta_G^k \tag{3.8a}$$

and

$$F_{G1}(\underset{\sim}{\theta}) = \sum_{\ell=0}^{f-1} \tilde{p}_{G\ell}\theta_A^\ell \quad . \tag{3.8b}$$

The argument used to construct the universal consistency relation, Eq. (2.17), for one-type processes can be generalized to yield

$$F_{A1}(\underset{\sim}{\theta}) = F'_{A0}(\underset{\sim}{\theta})/F'_{A0}(\underset{\sim}{1}) \tag{3.9a}$$

and

$$F_{G1}(\underset{\sim}{\theta}) = F'_{G0}(\underset{\sim}{\theta})/F'_{G0}(\underset{\sim}{1}) \quad , \tag{3.9b}$$

where the prime denotes differentiation with respect to the single component of θ present in the definition of $F_{A0}(\underset{\sim}{\theta}) = F_{A0}(\theta_G)$ and $F_{G0}(\underset{\sim}{\theta}) = F_{G0}(\theta_A)$. Thus,

$$\tilde{p}_{Ak} = M_A^{-1}(k+1)p_{A,k+1} \tag{3.10a}$$

and

$$\tilde{p}_{G\ell} = M_G^{-1}(\ell+1)p_{G,\ell+1} \quad , \tag{3.10b}$$

where M_A and M_G are the mean number of offspring from a parent of type A and type G, respectively, in the zeroth generation, i.e.,

$$M_A = \sum_{k=0}^{1} (k+1) p_{A,k+1} = F'_{A0}(1) \quad , \tag{3.11a}$$

$$M_G = \sum_{\ell=0}^{f-1} (\ell+1) p_{G,\ell+1} = F'_{G0}(1) \quad . \tag{3.11b}$$

The beauty of relations (3.9) is that now the generating functions $\underset{\sim}{F}_0(\underset{\sim}{\theta})$ and $\underset{\sim}{F}_1(\underset{\sim}{\theta})$ are defined in terms of a reduced set of parameters, $\{p_{Ak}; k = 0, 1, 2\}$, the probability of an antibody having k sites bound; and $\{p_{G\ell}; \ell = 0, 1, \ldots, f\}$, the probability of an antigen having ℓ sites bound. From these two generating functions we can calculate weight fractions, but first we formally represent the relationship between weight fractions and probability distributions for trees composed of antigens and antibodies.

3.4 Binomial and Multinomial Coefficients

Throughout the remainder of this monograph we shall utilize the following properties of binomial and multinomial coefficients. By definition (Feller, 1968), for any integers n and i

$$\binom{n}{i} = \begin{cases} \dfrac{n!}{i!(n-i)!} & , \quad 0 \leq i \leq n \\ 0 & , \quad \text{otherwise} \end{cases} \tag{3.12}$$

where $0! = 1$. Note, if $i > n$, or if n or i are negative, then the binomial coefficient is defined to be zero. As can easily be verified from the definition,

$$i\binom{n}{i} = n\binom{n-1}{i-1} \quad . \tag{3.13}$$

The generalization of a binomial coefficient to a <u>multinomial coefficient</u> is apparent if we write $\underset{\sim}{i} = (i_1, i_2, \ldots, i_f)$ and define

$$\binom{n}{\underset{\sim}{i}} = \begin{cases} \dfrac{n!}{i_1! i_2! \ldots i_f!} & , \quad i_k \geq 0 \quad , \quad k = 1, 2, \ldots, f, \text{ and } \sum_{k=1}^{f} i_k = n \quad , \\ 0 & , \quad \text{otherwise} \quad . \end{cases} \tag{3.14}$$

When f = 2, the multinomial coefficient reduces to the binomial coefficient.

C. WEIGHT FRACTIONS

__3.5__ Let

$w_{ij}^{(A)}$ = P(a rooted tree contains i particles of type A and j particles of type G, given the root is of type A) ,

$w_{ij}^{(G)}$ = P(a rooted tree contains i particles of type A and j particles of type G, given the root is of type G)

and

ρ = P(randomly chosen root is of type A) .

Then the probability, w_{ij}, of a rooted tree containing i antibodies and j antigens is

$$w_{ij} = \rho w_{ij}^{(A)} + (1 - \rho) w_{ij}^{(G)} . \qquad (3.15)$$

In identifying an aggregate with a branching process, each particle in the aggregate is chosen as the root of one family tree. Thus in a system containing a total of A antibodies and G antigens, the probability that a randomly selected particle chosen to be a root is an antibody, $\rho = A/(A + G)$.

__3.6 Remark__ In what follows we call w_{ij} the weight fraction of aggregates containing i antibodies and j antigens. However, in analogy with the arguments made in Section B of Chapter 2, one can see that this is rigorously true only if antibody and antigen each have unit weight. If one wishes to calculate a proper weight fraction for a system in which antibody and antigen have molecular weights MW_A and MW_G, respectively, then as Gordon (1962) and Gordon and Malcomb (1966) have shown, choosing ρ as the proper weight fraction of antibody in the system [i.e. $\rho = MW_A A/(MW_A A + MW_G G)$] will allow proper weight fractions to be computed with Eq. (3.15). We prove this result in Section 3.12.

<u>3.7</u> To compute weight fractions we must determine $w_{ij}^{(A)}$ and $w_{ij}^{(G)}$. These quantities can be expressed in terms of $\{p_{Ak}; k = 0, 1, 2\}$ and $\{p_{G\ell}; \ell = 0, 1, \ldots, f\}$ using a theorem from Good (1960), which is a multidimensional extension of Theorem 1. Before stating this result, given as Theorem 2 below, we need to introduce some additional notation.

Define the weight fraction generating functions of antigen-antibody aggregates by

$$\underline{W}(\underline{\theta}) = (W_A(\underline{\theta}), W_G(\underline{\theta}))$$

$$= \left(\sum_i \sum_j w_{ij}^{(A)} \theta_A^i \theta_G^j, \sum_i \sum_j w_{ij}^{(G)} \theta_A^i \theta_G^j \right) \quad , \tag{3.16}$$

where $\underline{\theta} = (\theta_A, \theta_G)$ is a dummy variable and $|\theta_i| \leq 1$, $i = A, G$. Then as Good (1955) and Gordon (1962) have shown

$$\underline{W}(\underline{\theta}) = \underline{\theta} \otimes \underline{F}_0(\underline{\theta} \otimes \underline{F}_1(\underline{\theta} \otimes \underline{F}_1(\ldots(\underline{\theta})\ldots))) \quad , \tag{3.17}$$
$$|\text{———}\infty \text{ times ———}|$$

where the direct product $\underline{\theta} \otimes \underline{F}_1(\underline{\theta})$ is defined by

$$\underline{\theta} \otimes \underline{F}_1(\underline{\theta}) \equiv (\theta_A F_{A1}(\underline{\theta}), \theta_G F_{G1}(\underline{\theta})) \quad .$$

$\underline{W}(\underline{\theta})$ can be expressed in a more compact form. Defining $\underline{u}(\underline{\theta}) = (u_A(\underline{\theta}), u_G(\underline{\theta}))$ by

$$\underline{u}(\underline{\theta}) = \underline{\theta} \otimes \underline{F}_1(\underline{\theta} \otimes \underline{F}_1(\ldots(\underline{\theta})\ldots)) \quad ,$$
$$|\text{———}\infty \text{ times ———}|$$

we see

$$\underline{W}(\underline{\theta}) = \underline{\theta} \otimes \underline{F}_0(\underline{u}(\underline{\theta})) \quad , \tag{3.18}$$

where

$$\underline{u}(\underline{\theta}) = \underline{\theta} \circledast \underline{F}_1(\underline{u}(\underline{\theta})) \quad . \tag{3.19}$$

3.8 Remark Equations (3.18) and (3.19) are simply the multidimensional analogues of Eqs. (2.24) and (2.25).

3.9 A difficulty lies in extracting $w_{ij}^{(A)}$ and $w_{ij}^{(G)}$ from Eq. (3.18) since Eq. (3.19) must first be solved for $\underline{u}(\underline{\theta})$. A theorem that we modify from Good (1960, 1965) exploits the implied relationship between $\underline{u}$ and $\underline{\theta}$ to obtain the desired coefficients. As a preliminary, observe from the definition of $\underline{W}(\underline{\theta})$ that

$$w_{ij}^{(A)} = C(\theta_A^i \theta_G^j)\{W_A(\underline{\theta})\} \tag{3.20a}$$

and

$$w_{ij}^{(G)} = C(\theta_A^i \theta_G^j)\{W_G(\underline{\theta})\} \quad . \tag{3.20b}$$

Theorem 2 Assuming $F_{A1}(\underline{0}) \neq 0$ and $F_{G1}(\underline{0}) \neq 0$,

$$w_{ij}^{(A)} = \begin{cases} \dfrac{M_A}{(i-1)j} C(u_A^{i-2}) \left\{ \dfrac{d[F_{G1}^j]}{du_A} \right\} C(u_G^{j-1})\{F_{A1}^i\} & , \quad i \geq 2 \, , \quad j \geq 1 \\ \\ 0 & , \quad i \geq 2 \, , \quad j = 0 \\ \\ C(u_A^0)\{F_{G1}^j(\underline{u})\} C(u_G^j)\{F_{A0}(\underline{u})\} & , \quad i = 1 \, , \quad j \geq 0 \\ \\ 0 & , \quad i = 0 \, , \quad j \geq 0 \end{cases} \tag{3.21a}$$

and

$$w_{ij}^{(G)} = \begin{cases} \dfrac{M_G}{i(j-1)} \, C(u_A^{i-1})\{F_{G1}^j\} C(u_G^{j-2}) \left\{\dfrac{d[F_{A1}^i]}{du_G}\right\} &, \; j \geq 2 \;,\; i \geq 1 \\[6pt] 0 &, \; j \geq 2 \;,\; i = 0 \\[6pt] C(u_A^i)\{F_{G0}(\underline{u})\} C(u_G^0)\{F_{A1}^i(\underline{u})\} &, \; j = 1 \;,\; i \geq 0 \\[6pt] 0 &, \; j = 0 \;,\; i \geq 0 \;, \end{cases} \quad (3.21b)$$

where $M_A = F'_{A0}(1)$ and $M_G = F'_{G0}(1)$. The proof of this theorem is given in Appendix A. Tutte (1975) gives a purely combinatorial proof of a more general multidimensional version of this theorem.

3.10 **Remark** The conditions $F_{A1}(\underline{0}) \neq 0$, $F_{G1}(\underline{0}) \neq 0$ will generally, although not necessarily, be met in systems restricted to finite-sized aggregates. From Eq. (3.8), $F_{A1}(\underline{0}) = \tilde{p}_{A0}$ and $F_{G1}(\underline{0}) = \tilde{p}_{G0}$. If $\tilde{p}_{A0} = 0$, then $\tilde{p}_{A1} = 1$, and each antibody in all generations after the first must have one offspring. Thus in trees which contain antibodies in addition to the root, all antibodies will act as connectors between antigens, and the process can be reduced to one involving antigens alone. If $\tilde{p}_{A0}$ and $\tilde{p}_{G0}$ are both zero, then in all generations after the first each antibody and each antigen must have at least one offspring and the family tree will be infinite. If $\tilde{p}_{G0} = 0$, but $\tilde{p}_{A0} \neq 0$, then the tree may be finite, and another method not dependent on Lagrange expansion would be required to find the weight fraction distribution.

3.11 **Example** Goldberg's (1952) Result

In order to clarify the operation of Theorem 2 we will apply it to antigen-antibody aggregate formation in solution and invoke the "equally reactive sites" hypothesis.

Let p_A be the probability that an antibody site is bound and p_G be the probability that an antigen site is bound. For a system containing A antibodies and G antigens

$$p_A = rp_G \tag{3.22}$$

where

$$r = fG/2A \tag{3.23}$$

is the antigen site-antibody site ratio. This fundamental relation obtains since each bound antibody site must be bound to an antigen site. Stochastic independence of sites, the probabilistic form of equal reactivity, implies

$$F_{A0}(\underline{u}) = (1-p_A + p_A u_G)^2 \quad , \quad F_{G0}(\underline{u}) = (1-p_G + p_G u_A)^f \tag{3.24a}$$

and

$$F_{A1}(\underline{u}) = (1-p_A + p_A u_G) \quad , \quad F_{G1}(\underline{u}) = (1-p_G + p_G u_A)^{f-1} \quad . \tag{3.24b}$$

By Eq. (3.18)

$$W_A(\underline{\theta}) = \theta_A(1-p_A + p_A u_G(\underline{\theta}))^2 \quad , \quad W_G(\underline{\theta}) = \theta_G(1-p_G + p_G u_A(\underline{\theta}))^f \quad . \tag{3.25}$$

It is worth emphasizing here that the assumption of stochastic independence reduces the number of independent parameters in $\underline{W}(\underline{\theta})$ from $(f + 2)$ to 2, i.e., p_A and p_G, and Eq. (3.22) further reduces it to 1, i.e., p_G.

To use Theorem 2 we need

$$M_A = F'_{A0}(1) = 2p_A \quad , \quad M_G = F'_{G0}(1) = fp_G \quad , \tag{3.26a}$$

$$\frac{d[F_{G1}^j]}{du_A} = jp_G(f-1)(1-p_G + p_G u_A)^{j(f-1)-1} \quad , \tag{3.26b}$$

and

$$\frac{d[F_{A1}^i]}{du_G} = ip_A(1-p_A + p_A u_G)^{i-1} \quad . \tag{3.26c}$$

Substituting into Eq. (3.21a) gives

$$w_{ij}^{(A)} = \frac{1}{i-1} [2p_A p_G (f-1) C(u_A^{i-2})\{(1-p_G+p_G u_A)^{j(f-1)-1}\} C(u_G^{j-1})\{(1-p_A+p_A u_G)^i\}]$$

$$= \frac{2(f-1)}{i-1} \binom{j(f-1)-1}{i-2}\binom{i}{j-1} p_A^j (1-p_A)^{i-j+1} p_G^{i-1} (1-p_G)^{j(f-1)-i+1} ,$$
$$i \geq 2 , \quad j \geq 1 , \quad (3.27a)$$

$$w_{i0}^{(A)} = 0 \qquad\qquad , \quad i \geq 2 , \qquad (3.27b)$$

$$w_{1j}^{(A)} = (1-p_G)^{j(f-1)} \binom{2}{j}(1-p_A)^{2-j} p_A^j \qquad , \quad j \geq 0 , \qquad (3.27c)$$

and

$$w_{0j}^{(A)} = 0 \qquad\qquad , \quad j \geq 0 . \qquad (3.27d)$$

Similarly, substituting into Eq. (3.21b) gives

$$w_{ij}^{(G)} = \frac{1}{j-1} [p_A p_G^f C(u_A^{i-1})\{(1-p_G+p_G u_A)^{j(f-1)}\} C(u_G^{j-2})\{(1-p_A+p_A u_G)^{i-1}\}]$$

$$= \frac{f}{j-1} \binom{j(f-1)}{i-1}\binom{i-1}{j-2} p_A^{j-1} (1-p_A)^{i-j+1} p_G^i (1-p_G)^{j(f-1)-i+1} ,$$
$$j \geq 2 , \quad i \geq 1 , \quad (3.28a)$$

$$w_{0j}^{(G)} = 0 \qquad\qquad , \quad j \geq 2 , \qquad (3.28b)$$

$$w_{i1}^{(G)} = \binom{f}{i}(1-p_G)^{f-i} p_G^i (1-p_A)^i \qquad , \quad i \geq 0 , \qquad (3.28c)$$

and

$$w_{i0}^{(G)} = 0 \qquad\qquad , \quad i \geq 0 . \qquad (3.28d)$$

We combine $w_{ij}^{(A)}$ and $w_{ij}^{(G)}$ according to Eq. (3.15). Since a randomly chosen root can be any antibody or antigen,

$$\rho = \frac{A}{A+G} \qquad\qquad (3.29)$$

for a system containing A antibodies and G antigens. Using Eqs. (3.22) and (3.23) to write p_A in terms of p_G and substituting into Eq. (3.15) gives the weight fraction of (i,j)-mer in solution as

$$w_{ij} = \frac{fG}{A+G} \frac{(i+j)[j(f-1)]!}{j!(i-j+1)![j(f-1)-i+1]!} p_G^{i+j-1} r^{j-1} (1-p_G)^{j(f-1)-i+1} (1-rp_G)^{i-j+1} \quad ,$$
$$i + j > 0 \quad . \quad (3.30)$$

The now classical result due to Goldberg (1952) for the molecular concentration of (i,j)-mer, m_{ij}, follows from Eq. (3.30) by using the relationship

$$w_{ij} = \frac{(i+j)m_{ij}}{A+G} \quad . \quad (3.31)$$

Hence

$$m_{ij} = fG \frac{[j(f-1)]!}{j!(i-j+1)![j(f-1)-i+1]!} p_G^{i+j-1} r^{j-1} (1-p_G)^{j(f-1)-i+1} (1-rp_G)^{i-j+1} \quad ,$$
$$i + j > 0 \quad . \quad (3.32)$$

The power of the generating function approach to aggregation problems can be appreciated more fully when one contrasts the systematic process leading to Eq. (3.30) with the more heuristic approaches of Goldberg (1952) and Stockmayer (1943) in their endeavors to pin down elusive combinatorial coefficients.

3.12 In accordance with Remark 3.6, the proper weight fraction of an (i,j)-mer, which we shall denote by W_{ij}, can be computed by choosing ρ as the proper weight fraction of antibody. Alternatively, m_{ij} can be multiplied by $(iMW_A + jMW_G)/(MW_A A + MW_G G)$. Either approach gives the same answer. To see this we need a very important result that relates $w_{ij}^{(A)}$ with $w_{ij}^{(G)}$.

Theorem 3

$$Ajw_{ij}^{(A)} = Giw_{ij}^{(G)} \quad , \quad i + j > 0 \quad . \quad (3.33)$$

__Proof__ Let N_{ij} be the number of (i,j)-mers in the chemical system. Then, following the argument of Section B of Chapter 2, iN_{ij} will be the number of antibody rooted trees corresponding to (i,j)-mers. By definition,

$$w_{ij}^{(A)} = \frac{\text{number of (i,j)-trees with an antibody root}}{\text{number of trees with an antibody root}}$$

$$= \frac{iN_{ij}}{\sum_{k,\ell} kN_{k\ell}} = \frac{iN_{ij}}{A} \quad . \tag{3.34}$$

Similarly,

$$w_{ij}^{(G)} = \frac{\text{number of (i,j)-trees with an antigen root}}{\text{number of trees with an antigen root}} = \frac{jN_{ij}}{G} \quad . \tag{3.35}$$

Multiplying Eq. (3.34) by j and Eq. (3.35) by i and then rearranging, yields Eq. (3.33), thus proving the theorem.

We now show that W_{ij}, the proper weight fraction of (i,j)-mer, may be calculated by either

$$W_{ij} = \frac{iMW_A + jMW_G}{MW_A A + MW_G G} m_{ij} \tag{3.36}$$

or

$$W_{ij} = \frac{MW_A A}{MW_A A + MW_G G} w_{ij}^{(A)} + \frac{MW_G G}{MW_A A + MW_G G} w_{ij}^{(G)} \quad . \tag{3.37}$$

To see the equivalence of these formulae, recall from Eqs. (3.31) and (3.15) that

$$m_{ij} = \frac{A + G}{i + j} w_{ij} = \frac{A + G}{i + j} \left[\frac{A}{A + G} w_{ij}^{(A)} + \frac{G}{A + G} w_{ij}^{(G)} \right] \quad .$$

Hence Eq. (3.36) can be rewritten as

$$W_{ij} = \frac{iMW_A + jMW_G}{MW_A A + MW_G G}\left[\frac{A}{i+j} w_{ij}^{(A)} + \frac{G}{i+j} w_{ij}^{(G)}\right]$$

$$= \frac{MW_A(iAw_{ij}^{(A)} + iGw_{ij}^{(G)}) + MW_G(jAw_{ij}^{(A)} + jGw_{ij}^{(G)})}{(i+j)(MW_A A + MW_G G)} .$$

Using Theorem 3, this can be simplified to yield

$$W_{ij} = \frac{MW_A(iAw_{ij}^{(A)} + jAw_{ij}^{(A)}) + MW_G(iGw_{ij}^{(G)} + jGw_{ij}^{(G)})}{(i+j)(MW_A A + MW_G G)}$$

$$= \frac{MW_A A w_{ij}^{(A)} + MW_G G w_{ij}^{(G)}}{MW_A A + MW_G G} ,$$

which is the alternative form, Eq. (3.37).

3.13 <u>Remark</u> Using Theorem 3 reduces the effort of calculating w_{ij} because only one of $w_{ij}^{(A)}$ and $w_{ij}^{(G)}$ need be evaluated by extracting coefficients from the weight fraction generating function. Cases in which i or j are zero would still need to be handled separately since $w_{0j}^{(G)}$ and $w_{i0}^{(A)}$ cannot be evaluated by use of Theorem 3.

CHAPTER 4

AGGREGATE SIZE DISTRIBUTION ON A CELL SURFACE

A. <u>GENERAL CONSIDERATIONS</u>

<u>4.1</u> One of the many benefits of using branching processes to study aggregation phenomena is the ability to specify arbitrary functions for p_{Ak} and $p_{G\ell}$, the respective probabilities of k sites on an antibody and ℓ sites on an antigen being bound. In Example 3.11 we examined the consequences of the equal reactivity assumption and derived the aggregate size distribution for solution phase antigen-antibody reactions. In reactions involving antigen (ligand) and antibody on cell surfaces (receptors), not all antigen sites are equally reactive. To characterize the differences in reactivity between an antigen in solution and an antigen bound to a cell surface, a binding model such as that developed by Gandolfi, Giovenco, and Strom (1978), DeLisi (1980), and Perelson (1981) is needed in order to specify p_{Ak} and $p_{G\ell}$. We shall proceed in our study of cell surface aggregation by using Theorem 2 to derive a general expression for the weight fraction distribution w_{ij} in which p_{Ak} and $p_{G\ell}$ are unspecified parameters. Then we shall show how this general expression can be simplified into a useable form when p_{Ak} and $p_{G\ell}$ are given by the binding model of Perelson (1981).

<u>4.2</u> When antigen and antibody sites are not all equally reactive we define the zeroth generation probability generating functions as

$$F_{A0}(\underline{u}) = \sum_{k=0}^{2} p_{Ak} u_G^k \quad ; \quad F_{G0}(\underline{u}) = \sum_{k=0}^{f} p_{Gk} u_A^k \quad . \tag{4.1}$$

The "universal consistency relations" determine the generating functions for all later generations to be

$$F_{A1}(\underline{u}) = M_A^{-1} \sum_{k=1}^{2} k p_{Ak} u_G^{k-1} \quad ; \quad F_{G1}(\underline{u}) = M_G^{-1} \sum_{k=1}^{f} k p_{Gk} u_A^{k-1} \qquad (4.2)$$

where

$$M_A \equiv \sum_{k=1}^{2} k p_{Ak} \quad ; \quad M_G \equiv \sum_{k=1}^{f} k p_{Gk} \quad . \qquad (4.3)$$

To calculate $w_{ij}^{(A)}$ for $i \geq 2$ and $j \geq 1$ from Eq. (3.21a), we require $C(u_A^{i-2})\{d[F_{G1}^j]/du_A\}$ and $C(u_G^{j-1})\{F_{A1}^i\}$. To find the first coefficient, note

$$\frac{d[F_{G1}^j]}{du_A} = j\left(M_G^{-1} \sum_{k=1}^{f} k p_{Gk} u_A^{k-1}\right)^{j-1} M_G^{-1} \sum_{k=2}^{f} k(k-1) p_{Gk} u_A^{k-2} \quad .$$

Applying the multinomial theorem (cf. Abramowitz and Stegun, 1964; p. 823) to expand the first term, one finds

$$C(u_A^{i-2})\left\{\frac{d[F_{G1}^j]}{du_A}\right\} = \sum_{k=2}^{\min(i,f)} k(k-1)(M_G^{-1} p_{Gk}) \sum_{\underline{n}} \binom{j-1}{\underline{n}} \prod_{\ell=1}^{f} (M_G^{-1} \ell p_{G\ell})^{n_\ell} \quad , \qquad (4.4)$$

where $\underline{n} = (n_1, n_2, \ldots, n_f)$ must satisfy the conditions

(i) $n_\ell \geq 0$, $\ell = 1, 2, \ldots, f$;

(ii) $\sum_{\ell=1}^{f} n_\ell = j-1$, $j \geq 1$;

(iii) $\sum_{\ell=1}^{f} (\ell-1) n_\ell = i-k$ for $k = 2, 3, \ldots, \min(i,f)$, $i \geq 2$, (4.5)

and the multinomial coefficient

$$\binom{j}{\underline{n}} = \frac{j!}{n_1! n_2! \ldots n_f!} \quad . \qquad (4.6)$$

To find the second coefficient note $F_{A1}(\underline{u})$ is a binomial and hence

$$C(u_G^{j-1})\{F_{A1}^i\} = M_A^{-i}\binom{i}{j-1}p_{A1}^{i-j+1}(2p_{A2})^{j-1} \quad . \tag{4.7}$$

Substitution of Eqs. (4.4) and (4.7) into Eq. (3.21a) gives

$$w_{ij}^{(A)} = \begin{cases} \dfrac{2M_A^{-(i-1)}M_G^{-j}}{i-1}\binom{i}{j-1}p_{A1}^{i-j+1}(2p_{A2})^{j-1}\sum_{k=2}^{\min(i,f)}\sum_{\underset{\sim}{n}}\binom{j-1}{\underset{\sim}{n}}\binom{k}{2}p_{Gk}\prod_{\ell=1}^{f}(\ell p_{G\ell})^{n_\ell} , \\ \qquad\qquad\qquad\qquad\qquad\qquad\qquad\qquad\qquad\qquad\qquad\qquad i \geq 2 \ , \ j \geq 1 \ , \\ 0 \qquad\qquad\qquad\qquad\qquad\qquad\qquad\qquad\qquad\qquad\qquad, \ i \geq 2 \ , \ j = 0 \ , \\ p_{Aj}\left(M_G^{-1}p_{G1}\right)^j \qquad\qquad\qquad\qquad\qquad\qquad\qquad\qquad, \ i = 1 \ , \ j \geq 0 \ , \\ 0 \qquad\qquad\qquad\qquad\qquad\qquad\qquad\qquad\qquad\qquad\qquad, \ i = 0 \ , \ j \geq 0 \ . \end{cases} \tag{4.8}$$

Repeating the process for Eq. (3.21b) or using Theorem 3 produces an analogous equation

$$w_{ij}^{(G)} = \begin{cases} \dfrac{M_A^{-i}M_G^{-(j-1)}}{j-1}\binom{i-1}{j-2}p_{A1}^{i-j+1}(2p_{A2})^{j-1}\sum_{\underset{\sim}{m}}\binom{j}{\underset{\sim}{m}}\prod_{\ell=1}^{f}(\ell p_{G\ell})^{m_\ell} , \\ \qquad\qquad\qquad\qquad\qquad\qquad\qquad\qquad\qquad\qquad j \geq 2 \ , \ i \geq 1 \ , \\ 0 \qquad\qquad\qquad\qquad\qquad\qquad\qquad\qquad\qquad\qquad, \ j \geq 2 \ , \ i = 0 \ , \\ p_{Gi}\left(M_A^{-1}p_{A1}\right)^i \qquad\qquad\qquad\qquad\qquad\qquad\qquad, \ j = 1 \ , \ i \geq 0 \ , \\ 0 \qquad\qquad\qquad\qquad\qquad\qquad\qquad\qquad\qquad\qquad, \ j = 0 \ , \ i \geq 0 \ , \end{cases} \tag{4.9}$$

where $\underset{\sim}{m} = (m_1, m_2, \ldots, m_f)$ must satisfy the conditions

(i) $\quad m_\ell \geq 0 \ , \quad \ell = 1, 2, \ldots, f \quad ;$

(ii) $\quad \displaystyle\sum_{\ell=1}^{f} m_\ell = j \ , \quad j \geq 2 \quad ;$

(iii) $\quad \displaystyle\sum_{\ell=1}^{f} (\ell-1)m_\ell = i - 1 \ , \quad i \geq 1 \quad .$ \hfill (4.10)

Lastly, $w_{ij}^{(A)}$ and $w_{ij}^{(G)}$ are combined according to Eq. (3.15) to obtain the general expressions for w_{ij}.

B. BIVALENT ANTIGENS

<u>4.3</u> It is useful to consider bivalent antigen in reaction with bivalent antibody to see a particular form for w_{ij} and the constraints imposed by valences on aggregate composition. While the constraints arise from purely mathematical considerations, they embody logical physical considerations.

For $f = 2$, the vectors $\underline{m}$ and $\underline{n}$ in Eqs. (4.8) and (4.9) are uniquely determined by the constraints (4.5) and (4.10). In fact, from Eq. (4.5) (ii) and (iii)

$$n_1 + n_2 = j - 1 \quad \text{and} \quad n_2 = i - 2 \quad,$$

and hence

$$n_1 = j - i + 1 \quad \text{and} \quad n_2 = i - 2 \quad, \tag{4.11}$$

provided $i \geq 2$, and $j - i + 1 \geq 0$. Similarly, from Eq. (4.10) (ii) and (iii)

$$m_1 + m_2 = j \quad \text{and} \quad m_2 = i - 1 \quad,$$

and hence

$$m_1 = j - i + 1 \quad \text{and} \quad m_2 = i - 1 \quad, \tag{4.12}$$

provided $i \geq 1$, and $j - i + 1 \geq 0$. Using these values, Eqs. (4.8) and (4.9) become

$$w_{ij}^{(A)} = \begin{cases} \binom{i}{j-1}\binom{j}{i-1} \dfrac{M_A^{-(i-1)} M_G^{-j}}{j} p_{A1}^{i-j+1}(2p_{A2})^{j-1} p_{G1}^{j-i+1}(2p_{G2})^{i-1} \quad, \\ \qquad\qquad\qquad\qquad\qquad\qquad\qquad\qquad\qquad i \geq 2 \;,\; j \geq 1 \;, \\ 0 \qquad\qquad\qquad\qquad\qquad\qquad\qquad\quad,\; i \geq 2 \;,\; j = 0 \;, \\ p_{Aj}\left(M_G^{-1} p_{G1}\right)^j \qquad\qquad\qquad\qquad\quad,\; i = 1 \;,\; j \geq 0 \;, \\ 0 \qquad\qquad\qquad\qquad\qquad\qquad\qquad\quad,\; i = 0 \;,\; j \geq 0 \;, \end{cases} \tag{4.13}$$

and

$$w_{ij}^{(G)} = \begin{cases} \binom{i}{j-i}\binom{j}{i-1} \dfrac{M_A^{-(i-1)}M_G^{-(i-1)}}{i} p_{A1}^{i-j+1}(2p_{A2})^{j-1}p_{G1}^{j-i+1}(2p_{G2})^{i-1} , \\ \hspace{6cm} j \geq 2 , i \geq 1 , \\ 0 \hspace{5cm} , j \geq 2 , i = 0 , \\ p_{Gi}\left(M_A^{-1}p_{A1}\right)^i \hspace{3cm} , j = 1 , i \geq 0 , \\ 0 \hspace{5cm} , j = 0 , i \geq 0 . \end{cases} \quad (4.14)$$

Applying Eq. (3.15) gives

$$w_{10} = \rho p_{A0} , \quad w_{01} = (1-\rho)p_{G0} ,$$

and (4.15)

$$w_{ij} = \binom{i}{j-1}\binom{j}{i-1}\dfrac{M_A^{-i}M_G^{-j}}{ij} p_{A1}^{i-j+1}(2p_{A2})^{j-1}p_{G1}^{j-i+1}(2p_{G2})^{i-1}(\rho i M_A + (1-\rho)jM_G) ,$$
$$i \geq 1 , j \geq 1 .$$

From the properties of the binomial coefficient [Eq. (3.12)], $w_{ij} = 0$ if $|i-j| \geq 1$. Thus if there are i bivalent antibodies in an aggregate, there can only be i-1, i, or i+1 bivalent antigens. This is to be expected since aggregates of bivalent antigen and bivalent antibody must be linear chains with antigens and antibodies alternating along the chain.

4.4 Linear aggregates formed between bivalent receptors (cell surface immunoglobulin) and bivalent ligand have been well studied. Here we shall show that Eq. (4.15) reduces to a formula derived previously by Perelson and DeLisi (1980) if p_{Ak} and $p_{G\ell}$ are chosen in accordance with the binding model employed by them (see Fig. 4.1). Briefly, in this binding model it is assumed that a population of cells expressing bivalent receptors on their surfaces is suspended in a medium containing a total concentration C_0 of bivalent ligand. Let S_0 be the total concentration of receptor sites, i.e., twice the antibody (receptor) concentration, and $S(t)$ the concentration of free receptor sites at time t. Assume that at time t free antigen in solution at concentration C(t) binds to the surface via the interaction of a free antigen site with a free antibody site. Let $C_1(t)$ denote the concentration of antigen bound to the surface at only one of its sites. Further, assume such singly bound antigen can bind to a second receptor with its remaining free site to become a doubly bound antigen at concentration $C_2(t)$. By conservation of receptor sites

$$S_0 = S(t) + C_1(t) + 2C_2(t) \quad . \tag{4.16}$$

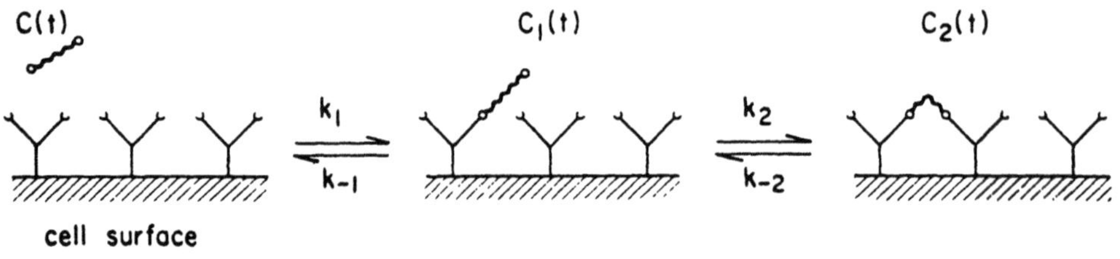

Figure 4.1

Perelson and DeLisi's (1980) model for bivalent antigen reversibly binding to bivalent cell surface receptors. Bivalent antigen free in solution at concentration C(t) reacts with free receptor sites present at concentration S(t). The concentration of antigen bound to the surface at exactly one site is denoted by $C_1(t)$. The fluidity of the membrane permits a singly bound antigen to react with a free receptor site on a second receptor. The concentration of antigen that cross-links two receptors is denoted by $C_2(t)$. The total concentration of antigen in the system $C_0 = C(t) + C_1(t) + C_2(t)$.

Using this notation, the probability that an antigen has one or two sites bound is given by

$$p_{G1} = C_1/C_0 \quad , \quad p_{G2} = C_2/C_0 \tag{4.17}$$

and hence from Eqs. (4.3), (4.16) and (4.17)

$$M_G = p_{G1} + 2p_{G2} = \frac{S_0 - S}{C_0} \quad . \tag{4.18}$$

In the model, receptor sites are assumed to act independently. If

$$p = \frac{S_0 - S}{S_0} \tag{4.19}$$

is used to denote the probability that a receptor site is bound, then the probability that k sites on a single receptor are bound is given by [cf. Eq. (1.1)]

$$p_{Ak} = \binom{2}{k}(1-p)^{2-k}p^k \quad , \quad k = 0, 1, 2 \tag{4.20}$$

and hence

$$M_A = p_{A1} + 2p_{A2} = 2p \quad . \tag{4.21}$$

Since there are $S_0/2$ antibody molecules and C_0 antigen molecules per unit volume, the probability, ρ, that an antibody is chosen as the root of a family tree is given by

$$\rho = \frac{S_0}{S_0 + 2C_0} \quad . \tag{4.22}$$

The substitution of these relations into Eq. (4.15) gives, after some straightforward algebra, a result which agrees with Eq. (81) in Perelson and DeLisi (1980). To make this comparison easier, let s be the number of singly bound antigens in a chain. In a linear chain with $i \geq 1$ antibodies, there are $i-1$ doubly bound antigens, and $s = 0, 1,$ or 2 singly bound antigens, depending upon how the chain terminates. The total number of antigens in the chain $j = i - 1 + s$. Writing Eq. (4.15) in terms of s, one finds

$$w_{is} = \frac{1}{2}\binom{2}{s}(M_A^{-1}p_{A1})^{2-s}(2M_A^{-1}p_{A2})^{i+s-2}(M_G^{-1}p_{G1})^s(2M_G^{-1}p_{G2})^{i-1}[\rho i M_A + (1-\rho)(i-1+s)M_G],$$
$$i \geq 1 \quad . \tag{4.23}$$

Substituting Eqs. (4.16)-(4.22) into Eq. (4.23) yields

$$w_{is} = \binom{2}{s} \frac{(2i-1+s)S_0}{S_0 + 2C_0} \left(\frac{S}{S_0}\right)^{2-s} \left(\frac{C_1}{S_0}\right)^s \left(\frac{2C_2}{S_0}\right)^{i-1} \quad , \quad i \geq 1 \quad . \tag{4.24}$$

Finally, introducing the relationship between weight fraction and concentration

$$w_{is} = \frac{(2i-1+s)m_{is}}{S_0/2 + C_0} \quad , \tag{4.25}$$

one obtains Eq. (81) of Perelson and DeLisi (1980):

$$m_{is} = \frac{S_0}{2} \binom{2}{s} \left(\frac{S}{S_0}\right)^{2-s} \left(\frac{C_1}{S_0}\right)^s \left(\frac{2C_2}{S_0}\right)^{i-1} \quad , \quad i \geq 1 \quad . \tag{4.26}$$

Perelson and DeLisi (1980) construct a system of differential equations based on the law of mass-action from which one can derive C_1 and C_2 as explicit functions of time. $S(t)$ is determined by Eq. (4.16), and the evolution in time of the aggregate size distribution can then be found from Eq. (4.24) or Eq. (4.26). We refer the reader to Perelson and DeLisi (1980) for details. The equilibrium size distribution can be obtained from Eq. (4.44) derived below for $f \geq 2$.

C. MULTIVALENT ANTIGENS

4.5 Here we present the major new result of the paper, the weight fraction distribution for aggregates formed when f-valent antigens bind and cross-link bivalent cell surface receptors. This result is directly applicable to the production of antibody by antigen-stimulated B lymphocytes, and to the release of histamine by basophils and mast cells. As for the case of bivalent antigens, we shall use a specific binding model to evaluate p_{Ak} and $p_{G\ell}$, and hence w_{ij}. Here we shall rely upon the model of Perelson (1981) shown schematically in Fig. 4.2. In this model antigen molecules, each with a total of v binding sites, are placed in the medium at concentration C_0 at time $t = 0$. Subsequently, antigen free in solution at concentration C reversibly binds to the surface to become singly bound antigen with concentration C_1. In this binding reaction any of the v antigen sites can bind with rate constant k_1 to a free receptor site present at concentration S. Once antigen is bound at a single site, an antigen-receptor complex is formed

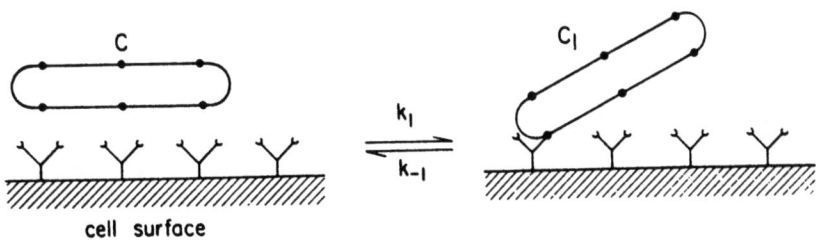

Figure 4.2

A model for multivalent antigens binding to and cross-linking bivalent cell surface receptors, taken from Perelson (1981). Antigen, at concentration C in the solution surrounding a population of cells, can bind at any one of v (v=6) sites to a free cell surface receptor site, with rate constant k_1. Singly bound antigen, at concentration C_1, can dissociate with rate constant k_{-1}, or bind -- using any of f-1 (f=3) free determinants -- a second receptor, with rate constant k_2. Doubly bound antigen, at concentration C_2, can return to the singly bound state by having either of the two receptor-antigen bonds break with rate constant k_{-2}, or can become triply bound with rate constant k_3. Triply bound antigen, at concentration C_3, can form no additional bonds with surface receptors, but any of its three bound sites can dissociate with rate constant k_{-3}.

which can dissociate with rate constant k_{-1}. By successive reversible binding reactions between free receptor sites and free antigen sites with forward rate constants k_2, k_3, ..., k_f and reverse rate constants k_{-2}, k_{-3}, ..., k_{-f}, the antigen becomes doubly bound, triply bound, etc. We denote by C_i the concentration of antigen bound to the surface at $i \geq 1$ of its functional groups.

In constructing equations to describe the model, we assume that once the antigen is confined to the surface (i.e., singly bound), not all of its free functional groups may be available for binding. We introduce the <u>effective valence</u>, f, to denote the total number of binding sites that can simultaneously bind to receptors (see Fig. 4.2). Hence only f-1 of the v-1 sites on a singly bound antigen are available for reaction with surface receptors. One could further assume, although we shall not do so here, that once the antigen is doubly bound, fewer than f-2 sites remain available for subsequent binding.

Assuming that receptor sites act independently and that all reactions are intermolecular, i.e., no cyclization reactions occur, then one can construct a system of differential equations for the binding kinetics. From these equations, one can show that at equilibrium (cf. Perelson, 1981; 1984)

$$C_i = \frac{KC}{f} \binom{f}{i} \kappa^{i-1} S^i \quad , \quad i = 1, 2, \ldots, f \quad , \tag{4.27}$$

where $K = vk_1/k_{-1}$ is the equilibrium constant for the binding of an antigen molecule to a single receptor site (i.e., v times the affinity of a receptor site for an antigen site) and

$$\kappa = \left[\prod_{j=2}^{i} k_j/k_{-j} \right]^{\frac{1}{i-1}} \tag{4.28}$$

is the geometric mean of the equilibrium constants for the second through the ith step in the cascade of reactions leading to the formation of an antigen with i sites bound. In the following development we assume that κ is a constant, i.e., independent of i.

Using this model we can explicitly evaluate p_{Ak} and $p_{G\ell}$. Since receptors are bivalent, with the receptor sites acting independently, p_{Ak} and M_A remain unchanged from the model described in Section 4.4 and are given by Eqs. (4.20) and (4.21). The probability that i sites on an antigen are bound is given by

$$p_{G0} = C/C_0 \quad , \tag{4.29a}$$

$$p_{Gi} = C_i/C_0 \quad , \quad i = 1, 2, \ldots, f \quad . \tag{4.29b}$$

By conservation of receptor sites

$$S_0 = S + \sum_{i=1}^{f} iC_i \tag{4.30}$$

and hence the mean number of sites bound on an antigen

$$M_G = \sum_{i=1}^{f} ip_{Gi} = (S_0 - S)/C_0 \quad , \tag{4.31}$$

as in the case of bivalent antigens.

Although one can solve a system of differential equations for $C_i(t)$ and hence obtain $p_{Gi}(t)$, no simplification of the multinomial coefficients in Eqs. (4.8) and (4.9) occur. Rather than pursue this approach, here we will restrict our attention to the equilibrium situation in which C_i is given by Eq. (4.27). The particular form of this expression permits considerable simplification of the combinatoric expressions in Eqs. (4.8) and (4.9), and a readily useable expression for the weight fraction distribution obtains.

In order to simplify the combinatorics, we need two results which we now state and prove.

4.6 <u>Lemma</u> For i, a positive integer, and j, a non-negative integer, let X_{ij} be the family of f-dimensional vectors $\underline{x} = (x_1, x_2, \ldots, x_f)$, $f \geq 2$, which satisfy the conditions

$$x_\ell \geq 0 \ , \quad \ell = 1, 2, \ldots, f \ , \tag{4.32}$$

$$\sum_{\ell=1}^{f} x_\ell = j \ , \tag{4.33}$$

and

$$\sum_{\ell=1}^{f} (\ell-1)x_\ell = i-1 \ . \tag{4.34}$$

Then

$$\sum_{X_{ij}} \binom{j}{\underset{\sim}{x}} \prod_{\ell=1}^{f} \binom{f-1}{\ell-1}^{x_\ell} = \binom{(f-1)j}{i-1} \ . \tag{4.35}$$

<u>Proof</u> We shall show that the left and right sides of Eq. (4.35) are the coefficients of a^{i-1} in two equivalent polynomial expansions of $(1+a)^{(f-1)j}$. Using the binomial expansion,

$$(1+a)^{(f-1)j} = \sum_{k=0}^{(f-1)j} \binom{(f-1)j}{k} a^k \ ,$$

one sees that the coefficient of a^{i-1} is the right side of Eq. (4.35). Alternatively, first using the binomial expansion of $(1+a)^{f-1}$ and then the multinomial expansion gives

$$(1+a)^{(f-1)j} = \left[\sum_{\ell=1}^{f} \binom{f-1}{\ell-1} a^{\ell-1}\right]^j$$

$$= \sum_{\underset{\sim}{x}} \binom{j}{\underset{\sim}{x}} \prod_{\ell=1}^{f} \binom{f-1}{\ell-1} a^{(\ell-1)x_\ell} \ ,$$

where the sum is taken over all $\underset{\sim}{x}$ that satisfy Eqs. (4.32) and (4.33). Rewriting the sum in terms of powers of a, one finds that the coefficient of a^{i-1} is the left side of Eq. (4.35) since the $\underset{\sim}{x}$ now also satisfy Eq. (4.34).

4.7 Lemma For integers i, j, k, f, with $j \geq 1$, $i \geq k$ and $2 \leq k \leq f$, let Y_{ijk} be the family of f-dimensional vectors $\underline{y} = (y_1, y_2, \ldots, y_f)$, which satisfy the conditions

$$y_\ell \geq 0, \quad \ell = 1, 2, \ldots, f, \tag{4.36}$$

$$\sum_{\ell=1}^{f} y_\ell = j-1, \tag{4.37}$$

and

$$\sum_{\ell=1}^{f} (\ell-1)y_\ell = i-k \quad \text{for} \quad k = 2, 3, \ldots, \min(i,f). \tag{4.38}$$

Then

$$\sum_{k=2}^{\min(i,f)} \binom{f}{k}\binom{k}{2} \sum_{Y_{ijk}} \binom{j-1}{\underline{y}} \prod_{\ell=1}^{f} \binom{f-1}{\ell-1}^{y_\ell} = \binom{f}{2}\binom{j(f-1)-1}{i-2}. \tag{4.39}$$

Proof

Use Lemma 4.6 to simplify the left side. Thus

$$\sum_{Y_{ijk}} \binom{j-1}{\underline{y}} \prod_{\ell=1}^{f} \binom{f-1}{\ell-1}^{y_\ell} = \binom{(f-1)(j-1)}{i-k}$$

and then

$$\sum_{k=2}^{f} \binom{f}{k}\binom{k}{2}\binom{(f-1)(j-1)}{i-k} = \binom{f}{2} \sum_{k=2}^{f} \binom{f-2}{k-2}\binom{(f-1)(j-1)}{i-k}$$

$$= \binom{f}{2} \sum_{k=2}^{f} C(a^{k-2})\{(1+a)^{f-2}\}C(a^{i-k})\{(1+a)^{(f-1)(j-1)}\}$$

$$= \binom{f}{2} C(a^{i-2})\{(1+a)^{f-2}(1+a)^{(f-1)(j-1)}\}$$

$$= \binom{f}{2}\binom{j(f-1)-1}{i-2}.$$

4.8 We are now in a position to compute $w_{ij}^{(A)}$ and $w_{ij}^{(G)}$. It is easy to verify that one can ignore the $i = 1$ case when computing $w_{ij}^{(A)}$ and the $j = 1$ case when computing $w_{ij}^{(G)}$ and yet still obtain a formula for w_{ij} valid for all positive i and j.

We first compute $w_{ij}^{(A)}$, $i \geq 2$. The term involving multinomials in Eq. (4.8) is simplified by substituting for p_{Gk} and $p_{G\ell}$ from Eqs. (4.27) and (4.29), to obtain

$$\sum_{k=2}^{\min(i,f)} \sum_{\underset{\sim}{n}} \binom{j-1}{\underset{\sim}{n}}\binom{k}{2} p_{Gk} \prod_{\ell=1}^{f} (\ell p_{G\ell})^{n_\ell}$$

$$= \left(\frac{KC}{fKC_0}\right)^j (\kappa S)^{i+j-1} f^{j-1} \sum_{k=2}^{\min(i,f)} \binom{f}{k}\binom{k}{2} \sum_{\underset{\sim}{n}} \binom{j-1}{\underset{\sim}{n}} \prod_{\ell=1}^{f} \binom{f-1}{\ell-1}^{n_\ell} ,$$

which by Lemma 4.7 reduces to

$$= \frac{f-1}{2}\left(\frac{KCS}{C_0}\right)^j (\kappa S)^{i-1} \binom{j(f-1)-1}{i-2} .$$

Thus Eq. (4.8) becomes

$$w_{ij}^{(A)} = \left(\frac{f-1}{i-1}\right)\binom{i}{j-1}\binom{j(f-1)-1}{i-2}(M_A^{-1} p_{A1})^{i-j+1}(2M_A^{-1} p_{A2})^{j-1} M_A \left(\frac{M_G^{-1} KCS}{C_0}\right)^j (\kappa S)^{i-1} ,$$

$$i \geq 2 , j \geq 1 .$$

Substituting the values of M_A, M_G, p_{A1}, p_{A2} and p from Eqs. (4.19)-(4.21) and (4.31) leads to

$$w_{ij}^{(A)} = 2\left(\frac{f-1}{i-1}\right)\binom{i}{j-1}\binom{j(f-1)-1}{i-2}(KC)^j(\kappa S_0)^{i-1}\left(\frac{S}{S_0}\right)^{2i} , \quad i \geq 2 , \quad j \geq 1 . \quad (4.40)$$

Using Lemma 4.6 in a similar set of manipulations, Eq. (4.9) can be reduced to

$$w_{ij}^{(G)} = \frac{1}{j-1}\binom{i-1}{j-2}\binom{j(f-1)}{i-1}(KC)^j(\kappa S_0)^{i-1}\left(\frac{S}{S_0}\right)^{2i}\left(\frac{S_0}{C_0}\right) , \quad j \geq 2 , \quad i \geq 1 . \quad (4.41)$$

With ρ given in Eq. (4.22), the weight fraction of aggregates with i antibodies and j antigens is

$$w_{10} = \frac{S_0}{S_0 + 2C_0} \left(\frac{S}{S_0}\right)^2 \quad,$$

$$w_{01} = \frac{2C_0}{S_0 + 2C_0} \left(\frac{C}{C_0}\right) \quad,$$

and (4.42)

$$w_{ij} = \frac{2S_0}{S_0 + 2C_0} \binom{i}{j-1}\binom{j(f-1)}{i-1} (KC)^j (\kappa S_0)^{i-1} \left(\frac{S}{S_0}\right)^{2i} \left(\frac{1}{i} + \frac{1}{j}\right) \quad, \quad i \geq 1 \quad, \quad j \geq 1 \quad.$$

4.9 Remark In order that w_{ij} be positive, it is necessary to have

$$j-1 \leq i \leq j(f-1) + 1 \quad, \quad \text{for } i \geq 1 \text{ and } j \geq 1 \quad.$$

The upper bound on the number of antibodies corresponds to the state in which every antigen site in the aggregate has reacted with a surface-bound antibody. The lower bound corresponds to the state in which the aggregate is a linear chain.

4.10 Using the relationship

$$w_{ij} = \frac{(i+j)m_{ij}}{S_0/2 + C_0} \tag{4.43}$$

between weight fraction and concentration, one finds

$$m_{10} = \frac{S_0}{2} \left(\frac{S}{S_0}\right)^2 \quad,$$

$$m_{01} = C \quad,$$

and (4.44)

$$m_{ij} = \frac{S_0}{ij} \binom{i}{j-1}\binom{j(f-1)}{i-1} (KC)^j (\kappa S_0)^{i-1} \left(\frac{S}{S_0}\right)^{2i} \quad , \quad i \geq 1 \; , \; j \geq 1 \; .$$

This, a practical result of great importance, gives the equilibrium concentration of cell surface aggregates in terms of the receptor affinity for antigen, K; the ligand valence, f; the equilibrium concentration of ligand free in solution, C; the geometric mean of the equilibrium constants for cell surface cross-linking reactions, κ; the total concentration of cell surface receptors, $S_0/2$; and the equilibrium concentration of free receptor sites, S. The equilibrium value of S must be obtained by solving the conservation equation (4.30) which, upon substitution of Eq. (4.27), takes the form

$$S_0 = S + KCS(1+\kappa S)^{f-1} \quad . \tag{4.45}$$

For $f > 2$, this equation is best solved numerically. The remaining parameters K, f, C, κ, and S_0 characterize the cell and ligand and must be estimated for each experimental situation (cf. Dembo, et al., 1979).

4.11 Our analysis to this point has involved reactions between f-valent antigen and bivalent antibody. Antibodies may have valence greater than two. For example, immunoglobulin M has valence ten. Non-immunoglobulin cell surface receptors may also have valences greater than two. Theorem 2 is completely general and not restricted to situations in which antibody is bivalent. In Appendix B we apply Theorem 2 to derive the weight fraction distribution for aggregates composed of g-valent antibody and f-valent antigen. When both the antibody sites and the antigen sites are equally reactive, we regain a result originally derived by Goldberg (1953). When the antigen sites are not equally reactive, for example when p_{Gk} is given by the model of Perelson (1981) [cf. Eqs. (4.27)-(4.29)], we obtain new results relevant to the binding and cross-linking of g-valent cell surface receptors. The results in Appendix B have been published separately in abbreviated form (Macken and Perelson, 1982).

CHAPTER 5

GELATION AND INFINITE-SIZED TREES

A. GENERAL CONSIDERATIONS

5.1 To this point our work has been predicated upon the assumption that all aggregates, whether composed of one or two types of particles, are finite-sized. With multivalent particles, infinite-sized aggregates may form (in idealized systems containing an infinite number of particles). Thus four questions of interest arise: 1) Under what conditions do aggregates grow without bound? 2) How is the total mass in the system apportioned between finite-sized and infinite-sized aggregates? 3) What is the weight fraction distribution of finite-sized aggregates in systems which also contain infinite-sized aggregates? and 4) How does the weight fraction distribution change in large but finite systems when infinite-sized aggregates are predicted to appear? Consideration of these questions has appeared in the polymer chemistry literature (cf. Stockmayer, 1943; Flory, 1953; Dobson and Gordon, 1964; Donoghue and Gibbs, 1978, 1979; Dusek, 1979; Ziff and Stell, 1980; Perelson and Goldstein, 1985) where infinite-sized aggregates are referred to as the gel, and finite-sized aggregates as the sol. Here we shall illustrate how the branching process approach can be utilized to answer the first three questions for antigen-antibody reactions. The effects on antigen-antibody aggregation of having only a finite number of molecules available for reaction will be dealt with in a separate publication (Malakoff and Perelson, 1985).

In extending our analysis to the gel phase, one needs to reconsider the assumption held until now that the effects of ring formation can be ignored. Consider the interaction of a free site on a sol phase aggregate with another free site either on the same aggregate or on another aggregate. Because there are only a finite number, n_s, of free sites on any sol aggregate, the ratio of n_s to the number of free sites on all other aggregates goes to zero as N, the number of aggregates in the system, goes to infinity. Hence if all free sites in the system are capable of interacting with each other, the probability of ring formation goes to zero as N goes to infinity. Falk and Thomas (1974) demonstrated this effect by comparing the results of Monte Carlo simulations in which rings were allowed, to simulations in which rings were forbidden. For large enough systems, both simulations gave essentially the same results. This situation changes once a gel forms, for now there can be an infinite number of free sites available for ring closure within the infinite-sized aggregate comprising the gel. In polymer chemistry, intramolecular reactions are believed to be prevalent in the gel (cf. Stockmayer, 1943; Gordon and Scantlebury, 1967; Gordon and Parker, 1970/71). The inclusion of intramolecular reactions is also necessary to avoid certain theoretical paradoxes such as the so-called Malthusian packing paradox which states that a tree-like molecule growing infinitely large could not fit into three-dimensional space (Gordon and Parker, 1970/71; Gordon and Ross-Murphy, 1975). Stepto (1982) provides a comprehensive review of recent results on intramolecular reactions within the context of polymer theory.

In order to incorporate rings into a gel phase we shall follow a procedure devised by Flory in which one postulates that the basic equations developed for a pre-gel system hold after the gel forms. Thus the generating functions and site occupancy probabilities are assumed to be the same in the pre-gel and post-gel states. Under these assumptions, one might expect that gelation has no effect on earlier analyses, but in fact we find otherwise. For example, in the random condensation of f-valent particles, one would retain Eq. (2.34) for w_n, the weight fraction of n-mer. However, in the post-gel regime (characterized by p being larger than some critical value, p_c) one finds that the weight fraction distribution is defective in the sense that $\Sigma w_n < 1$; the remaining weight being attributed to an infinite-sized aggregate, the gel. To see that Flory's procedure implicitly incorporates rings, note from Eq. (2.34) that as p increases past p_c, the weight in the gel increases. In the limit when p = 1, the system contains no sol phase aggregates (i.e., $w_n = 0$ for all n). However, if p = 1 then the gel contains no free sites, and hence all possible rings must have formed. Stockmayer (1943) and Ziff and Stell (1980) show in a more general and rigorous manner that the Flory procedure predicts the occurrence of intramolecular reactions in the gel for all values of $p > p_c$. An alternative procedure for treating rings in the gel, called the spanning-tree approximation, developed by Gordon and Scantlebury (1966, 1967), may also be followed.

<u>5.2</u> In this chapter, starting first with aggregates of one particle type whose sites are not necessarily equally reactive (i.e., offspring probabilities are not necessarily independent), we derive a general condition for gelation. Flory's (1941b) result will be seen as a special case. Next, we derive the general condition for gelation in antigen-antibody systems. When antibody sites and antigen sites are each assumed to be equally reactive, Goldberg's (1952) result is obtained. When the ligand is trivalent and the antibody a cell surface receptor, we recover the recent results of Goldstein and Perelson (1984). We find that our general condition for gelation, when applied to cell surface reactions involving multivalent ligands, predicts the appearance of infinite-sized receptor-ligand aggregates for certain "critical" parameter ranges, a prediction which may have practical implications. For example, the phenomenon of "receptor patching" seen on a wide variety of cells (cf. Taylor, et al., 1971; DeLisi and Perelson, 1976; Schreiner and Unanue, 1976) may in fact be a form of gelation. "Receptor patching" describes extensive binding of multivalent ligands to mobile cell surface receptors, leading to the formation of macroscopic aggregates. If the ligand is fluorescently labeled, then discrete fluorescent "patches" are seen to form on the cell surface. Yet another phenomenon that may be explained by the formation of gel-like cell surface aggregates was recently described by Peacock and Barisas (1983), who observed experimentally that the mobility of receptor-bound antigen decreased dramatically at critical values of the antigen concentration and valence.

B. GELATION IN A ONE-TYPE PROCESS

<u>5.3</u> We employ the notation of Chapter 2 to describe aggregation of f-valent particles. Let p_k be the probability that a zeroth generation particle reacts at k sites, k = 0, 1, ..., f; $\tilde{p}_k$ be the probability that a first or later generation particle reacts at k sites, k = 0, 1, ..., f-1; and Z_r be the number of particles in the rth generation. By the universal consistency relations $\tilde{p}_k = (k+1)p_{k+1}/m$, where m is the mean number of sites bound on a particle in the zeroth generation, i.e.,

$$m = \sum_{k=1}^{f} kp_k \quad . \tag{5.1a}$$

Further define

$$\tilde{m} = \sum_{k=1}^{f-1} k\tilde{p}_k \quad , \tag{5.1b}$$

the mean number of sites bound on a particle in the first or later generation.

Conditions for gelation may be obtained by considering the manner in which an aggregate is expected to grow with successive generations. To this end, we introduce the notation E(X) for the expected value of a random variable X, and E(X|Y) for the expected value of X, given the value of the random variable Y. Then the expected number of particles in the first generation, given that there are Z_0 particles in the zeroth generation, is

$$E(Z_1|Z_0) = mZ_0 \tag{5.2}$$

and the expected number of particles in the (r+1)st generation, given that there are Z_r particles in the rth generation, is

$$E(Z_{r+1}|Z_r) = \tilde{m}Z_r \quad , \quad r = 1, 2, \ldots \quad . \tag{5.3}$$

Iterating Eq. (5.3) and incorporating Eq. (5.2), one finds

$$E(Z_{r+1}|Z_0) = \tilde{m}\tilde{m}^r Z_0 \quad . \tag{5.4}$$

If $\tilde{m} < 1$, then $E(Z_{r+1}|Z_0) < E(Z_r|Z_0)$, and the expected number of particles decreases with each succeeding generation. In this case, the aggregate must be finite in size. Conversely, if $\tilde{m} > 1$, then $E(Z_{r+1}|Z_0) > E(Z_r|Z_0)$ and the aggregate will have a positive probability of growing without bound. As Flory (1941a) argued, the case

$$\tilde{m} = 1 \tag{5.5}$$

represents the critical condition for incipient formation of an infinite-sized aggregate. Equation (5.5) is thus taken to be the condition for gelation in a one-type system.

5.4 An alternative derivation of the above condition for gelation uses a generating function approach. Recall that the weight fraction generating function $W(\theta) = \theta F_0(u(\theta))$, where $u(\theta) = F_1(u(\theta))$, can be written as $W(\theta) = \sum_{n=1}^{\infty} w_n \theta^n$. Thus $W(1) = \sum_{n=1}^{\infty} w_n$ is the total weight of finite-sized aggregates. This motivates our defining W_s, the weight fraction of the sol, and W_g, the weight fraction of the gel, as

$$W_s \equiv W(1) \tag{5.6a}$$

and

$$W_g \equiv 1 - W(1) \quad . \tag{5.6b}$$

A gel phase exists whenever $W(1) < 1$. To determine the conditions under which $W(1) < 1$, we define a quantity v, such that,

$$W(1) = F_0(v) \quad , \tag{5.7}$$

where

$$v \equiv u(1) = F_1(v) \quad . \tag{5.8}$$

Because $F_1(v)$ and $F_0(v)$ are generating functions of normalized probability distributions [see Eqs. (2.1) and (2.16)], $F_0(v)$ and $F_1(v)$ are monotonically increasing functions of v for $0 < v < 1$, and $F_1(1) = F_0(1) = 1$. Since $F_1(1) = 1$, one solution to Eq. (5.8) is $v = 1$. Therefore $W(1) < 1$ (i.e., a gel exists) if and only if another solution to Eq. (5.8) exists with $v < 1$. Good (1963), Harris (1963), Karlin and Taylor (1975), and others derive conditions for $v < 1$ to be a solution of Eq. (5.8). A brief derivation follows.

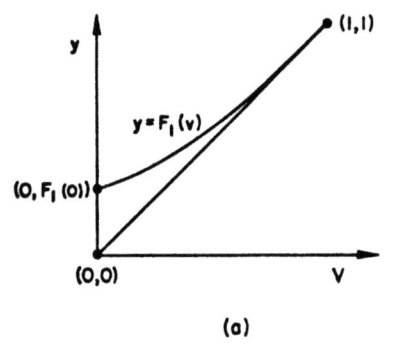 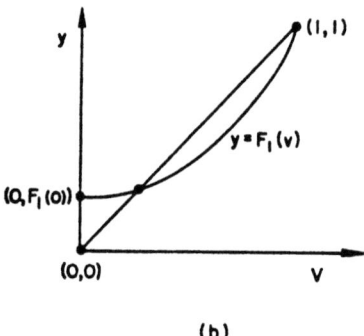

Figure 5.1

The graphs of $y = F_1(v)$ and $y = v$. (a) When $F_1'(1) \leq 1$, the curve $y = F_1(v)$ is above the line $y = v$ for $0 < v < 1$. (b) When $F_1'(1) > 1$, the curve $y = F_1(v)$ intersects the line $y = v$ at some value of $v < 1$.

Note that the power series for $F_1(v)$ has non-negative probabilities as coefficients, which ensures that $d^2F_1(v)/dv^2 \geq 0$. Assume this second derivative is strictly positive. (The case of a zero second derivative is treated in Remark 5.6.) Hence the graph of $y = F_1(v)$ is concave upward. Further, as shown in Fig. 5.1, the curve starts at the point $(0, F_1(0))$ above the line $y = v$, and ends at the point $(1,1)$ on the line $y = v$. Accordingly, only one of two possible situations can exist: Either (i) the curve $y = F_1(v)$ is entirely above the line $y = v$. In this case $v = 1$ is the unique root of Eq. (5.8). Further, as can be seen from

Fig. 5.1a or from the fact that $1 - F_1(v) \leq 1 - v$ for all v, the inequality $\tilde{m} = F_1'(1) \leq 1$ holds. Or (ii), the curve $y = F_1(v)$ intersects the line $y = v$ at some point $v < 1$ (see Fig. 5.1b). Since a curve that is concave upward intersects a straight line in at most two points, there are two solutions to Eq. (5.8): $v = 1$ and $v < 1$. Further, as can be seen from Fig. 5.1b, $\tilde{m} = F_1'(1) > 1$.

In summary, the solution $v = 1$ to Eq. (5.8) is unique if and only if $\tilde{m} \leq 1$, in which case $W(1) = 1$ and all aggregates are finite-sized. If $\tilde{m} > 1$, then Eq. (5.8) has a second solution, $v < 1$. This second solution is the relevant one for gelation since it is the smallest positive root of Eq. (5.8) (see Section 2.11). Thus when $\tilde{m} > 1$, $W(1) < 1$, implying the existence of infinite-sized aggregates.

5.5 Remark The quantity v can be interpreted as the probability that a subtree starting from an individual in the first or later generation goes extinct. Similarly, $W_s = F_0(v)$ can be interpreted as the probability that a tree starting from an individual in the zeroth generation goes extinct (Good, 1963).

5.6 Remark The proof that Eq. (5.8) has $v = 1$ as its unique solution if and only if $F_1'(1) \leq 1$, depends upon $F_1(v)$ having a positive second derivative. However for bivalent particles, $f = 2$, and by Eqs. (2.15) and (2.16), $F_1(v) = \tilde{p}_0 + (1 - \tilde{p}_0)v$. Thus $F_1''(v) = 0$. To treat this special case observe that if $0 \leq \tilde{p}_0 \leq 1$, then $F_1'(1) = 1 - \tilde{p}_0 \leq 1$. Simple algebra now shows that $v = 1$ is the only solution to Eq. (5.8). Thus the result that $v = 1$ is the only solution to Eq. (5.8) when $F_1'(1) \leq 1$ is still valid.

5.7 Remark When $\tilde{m} = 1$, $v = 1$, and the probability that a tree that reaches the first generation will go extinct is one. However, when $\tilde{m} = 1$, the expected size of an aggregate is infinite. Thus we arrive at a paradoxical situation in which the tree is certain to go extinct and yet its expected size is infinite. The condition $\tilde{m} = 1$ is thus a borderline situation for the occurrence of a gel. As we stated before, in polymer chemistry $\tilde{m} = 1$ is taken to be the <u>critical condition</u> for gelation.

5.8 Example Gelation Condition for f-valent Particles with Equally Reactive Sites

Assume particle sites, regardless of generation, are all equally reactive, with p being the probability that a site is bound (i.e., the extent of reaction). Then

$$\tilde{p}_k = \binom{f-1}{k} p^k (1-p)^{f-k-1}$$

and from Eq. (5.1b)

$$\tilde{m} = (f-1)p \quad .$$

The condition for gelation, $\tilde{m} = 1$, becomes

$$(f-1)p = 1 \quad , \tag{5.9}$$

a result originally derived by Flory (1941a).

C. GELATION IN A TWO-TYPE PROCESS (ANTIGEN-ANTIBODY REACTIONS)

5.9 We restrict our attention to antigen-antibody aggregates. This special class of two-type process is distinguished by the fact that antigens and antibodies alternate along any path through the aggregate. Consequently, a branching process which starts in generation zero with a single individual of type G will contain individuals solely of type G in generation 2n and individuals solely of type A in generation 2n + 1. If the expected number of antigens in generation 2n + 2 is greater than the expected number of antigens in generation 2n, then there is a positive probability that the aggregate will grow without bound. (An analogous argument can be applied to antibodies.) To state the argument formally, let Z_{ir} be the number of individuals of type i, i = A, G, in generation r; let

$$\tilde{M}_A \equiv \sum_{k=0}^{1} k \tilde{p}_{Ak} = F'_{A1}(1) \tag{5.10a}$$

be the mean number of offspring from an antibody parent in the first or later generation; and let

$$\widetilde{M}_G \equiv \sum_{k=0}^{f-1} k\widetilde{p}_{Gk} = F'_{G1}(1) \qquad (5.10b)$$

be the mean number of offspring from an antigen parent in the first or later generation. Then along the lines of the previous section, one can show (cf. Pollard, 1973; pp. 103-108) that for $r \geq 0$ and $Z_{i0} > 0$

$$E(Z_{i,2r+2}|Z_{i0}) > E(Z_{i,2r}|Z_{i0}) \quad , \quad i = A, G \qquad (5.11)$$

if and only if

$$\widetilde{M}_A \widetilde{M}_G > 1 \quad .$$

Thus the critical condition for gelation in antigen-antibody systems is

$$\widetilde{M}_A \widetilde{M}_G = 1 \quad . \qquad (5.12)$$

For a general two-type process Gordon (1962) derived an analogous relationship by examining the condition under which the mean of the weight fraction distribution diverges.

5.10 Example   Antibody and Antigen Sites Both Equally Reactive

Let $r = fG/2A$ be the ratio of antigen to antibody sites, p_G be the probability that an antigen site has reacted, and $p_A = rp_G$ be the probability that an antibody site has reacted. From Eqs. (5.10) and (3.24)

$$\widetilde{M}_A = p_A = rp_G \quad , \quad \widetilde{M}_G = (f-1)p_G \quad .$$

The condition for gelation, Eq. (5.12), becomes

$$(f-1)rp_G^2 = 1 \quad , \qquad (5.13)$$

a result of Goldberg's (1952) analysis of solution phase antigen-antibody reactions. Watson (1958) derived this equation directly from a branching process analysis of antigen-antibody reactions.

D. INFINITE-SIZED AGGREGATES ON A CELL SURFACE

5.11 As a new application of Eq. (5.12) we shall derive the condition for the formation of infinite-sized receptor-ligand aggregates on cell surfaces. The generation of such clusters is the two-dimensional analog of gelation in three-dimensional systems and for convenience will still be referred to as gelation. DeLisi and Perelson (1976), Perelson (1980), and Goldstein and Perelson (1984) have previously introduced the concept of cell surface gelation and have calculated for special cases or under limiting approximations the conditions for gelation.

To proceed, we adopt the model presented in Section 4.5 for the interaction of an f-valent antigen with bivalent cell surface receptors. Confining our attention to the equilibrium state, Eqs. (4.27) and (4.29) give

$$P_{G0} = \frac{C}{C_0} \quad ,$$

$$P_{Gi} = \frac{KC}{f\kappa C_0} \binom{f}{i} (\kappa S)^i \quad , \quad i = 1, 2, \ldots, f \quad . \tag{5.14}$$

Now, Eqs. (5.10b) and (3.10b) imply

$$\tilde{M}_G = M_G^{-1} \sum_{k=0}^{f-1} k(k+1) P_{G,k+1} \quad . \tag{5.15}$$

Substituting Eq. (5.14) into Eq. (5.15), performing the summation, and using Eq. (4.31) for M_G yields

$$\tilde{M}_G = \frac{KC(f-1)\kappa S^2 (1+\kappa S)^{f-2}}{S_0 - S} \quad . \tag{5.16}$$

As in Example 5.10, bivalent cell surface receptor sites are assumed to act independently, and hence $\tilde{M}_A$ equals the probability that an antibody (receptor) site is bound. Thus

$$\tilde{M}_A = \frac{S_0 - S}{S_0} \quad . \tag{5.17}$$

Substituting into Eq. (5.12), the gelation condition for this ligand binding model becomes

$$\tilde{M}_A \tilde{M}_G = \frac{KC(f-1)\kappa S^2(1+\kappa S)^{f-2}}{S_0} = 1 \quad , \tag{5.18a}$$

where S solves the conservation equation

$$S_0 = S + KCS(1 + \kappa S)^{f-1} \quad . \tag{5.18b}$$

5.12 Remark Goldstein and Perelson (1984) have independently derived Eq. (5.18) for the case of trivalent antigens using a totally different combinatorial method. Macken and Perelson (1982) give the appropriate generalization of Eq. (5.18) for the case of g-valent receptors.

5.13 Using Eq. (5.18), it is easy to show that infinite-sized aggregates on a cell surface are not possible when ligands and receptors are both bivalent. With $f = 2$, Eq. (5.18) becomes

$$\frac{KC\kappa S^2}{S_0} = 1 \quad , \tag{5.19a}$$

where S solves

$$S_0 = S + KCS(1 + \kappa S) \quad . \tag{5.19b}$$

Equation (5.19b) can be rearranged to give

$$\frac{KC\kappa S^2}{S_0} + \frac{(1 + KC)S}{S_0} = 1 \quad . \tag{5.20}$$

Since $(1 + KC)S/S_0 > 0$, Eq. (5.20) shows that $KC\kappa S^2/S_0 < 1$. Thus the condition for gelation, Eq. (5.19a), can never be met.

An even simpler argument for the impossibility of gelation is that bivalent receptors and ligands have at most one offspring in generations after the zeroth. Thus $\tilde{M}_A \leq 1$ and $\tilde{M}_G \leq 1$, implying that $\tilde{M}_A \tilde{M}_G \leq 1$. The critical condition can be satisfied but never exceeded. At the critical condition $W(1) = 1$, and thus the weight fraction of gel, $W_g = 1 - W(1)$, is zero.

<u>5.14</u>   The condition for gelation on a cell surface, Eq. (5.18), is important in applications. The results of Section 4.8 for the size distribution of cell surface aggregates are only valid in the pre-gel state where all aggregates have finite size. As parameter values in the binding and cross-linking model are changed, the condition for gelation may be met. Once there is a positive probability of a gel forming, the results of Section 4.8 are no longer valid. This breakdown of the model in certain parameter ranges was observed by Perelson (1981) but not directly attributed to gelation. Numerical studies (unpublished) have now verified that the parameter values for which the binding and cross-linking model gave physically unrealistic results all fall in regimes in which the gelation criterion is met.

5.15   The Critical Antigen Concentration for Gelation

We can rewrite Eqs. (5.18a) and (5.18b) in such a way that we can solve explicitly for the critical antigen concentration C^* required to reach the sol-gel transition. Write Eq. (5.18b) as

$$\frac{S_0 - S}{KCS(1 + \kappa S)} = (1 + \kappa S)^{f-2} \tag{5.21}$$

and substitute into Eq. (5.18a) to give

$$\tilde{M}_A \tilde{M}_G = \frac{(f - 1)\kappa S(S_0 - S)}{S_0(1 + \kappa S)} \quad . \tag{5.22}$$

At the gel point, $\tilde{M}_A\tilde{M}_G = 1$, and hence

$$(f - 1)\kappa S^2 - (f - 2)\kappa S_0 S + S_0 = 0 \quad . \tag{5.23}$$

Let S^* denote the value of S at the gel point. Solving Eq. (5.23), we find

$$S^* = \frac{(f - 2)S_0}{2(f - 1)}\left[1 \pm \sqrt{1 - \frac{2(f - 1)}{(f - 2)^2 \beta}}\right] \quad , \tag{5.24}$$

where $\beta = \kappa S_0/2$ is a dimensionless cross-linking constant. The factor of two in the denominator is introduced to conform with the usage in Goldstein and Perelson (1984) in which the cell surface receptor density, $S_0/2$, is used as a primary parameter. Because S is real, we require

$$\beta \geq \frac{2(f - 1)}{(f - 2)^2} \quad . \tag{5.25}$$

If condition (5.25) is not met, then gelation cannot occur. Hence, we immediately confirm our previous finding that gelation is impossible if $f = 2$.

Solving Eq. (5.18a) for C, we see that at the gel point

$$C^* = \frac{1}{2K(f - 1)\beta(S^*/S_0)^2(1 + 2\beta S^*/S_0)^{f-2}} \quad . \tag{5.26}$$

By substituting Eq. (5.24) for S^* into Eq. (5.26), we obtain an explicit value for the critical concentration of antigen required for gelation.

<u>5.16</u> <u>Example</u> The simplest example of a cell surface antigen-antibody system that gels is one in which the antigen is trivalent. With $f = 3$, we require

$$\beta \geq 4 \quad . \tag{5.27}$$

Further,

$$S^* = \frac{S_0}{4}\left[1 \pm \sqrt{1 - \frac{4}{\beta}}\right] \tag{5.28}$$

and

$$c^* = \frac{1}{4K\beta(S^*/S_0)^2(1 + 2\beta S^*/S_0)} \quad . \tag{5.29}$$

Substituting the two values of S^* into Eq. (5.29), we find that

$$c^*_\pm = \frac{2}{K[\beta^2 - 2\beta - 2 \mp \beta^{3/2}(\beta - 4)^{1/2}]} \quad , \tag{5.30}$$

where c^*_+ denotes the larger, and c^*_- the smaller, of the two values for c^*. When $\beta = 4$, $c^*_\pm = 1/(3K)$, and the two values of c^* coincide. When $\beta > 4$, the two values of c^* are distinct, and the gel phase exists for all values of C such that $c^*_- \leq C \leq c^*_+$. In Fig. 5.2 we show the locus of c^*_+ and c^*_- values for different values of β and indicate the sol and gel regions.

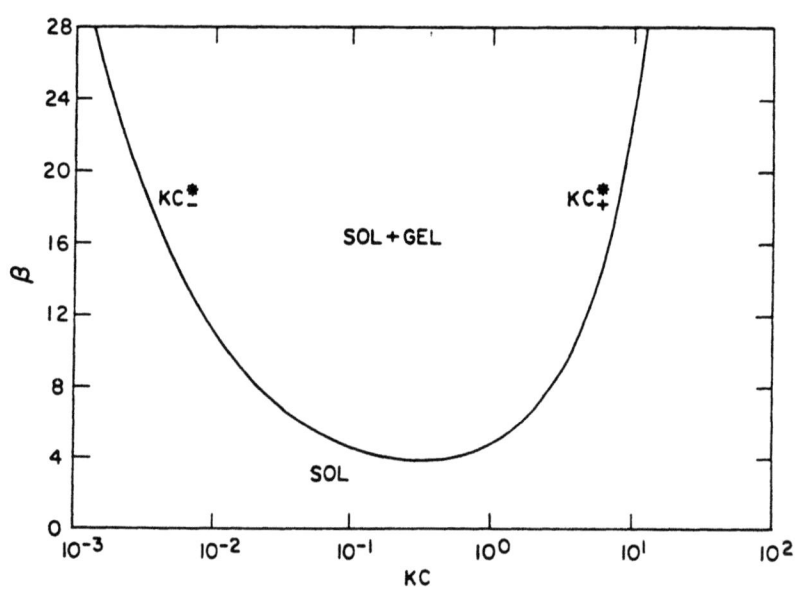

Figure 5.2

The solid line indicates the locus of KC^*_+ and KC^*_- values for different values of β. A gel exists only when $\beta > 4$ and for those values of C between c^*_- and c^*_+.

__5.17__ Although we worked Example 5.16 using $f = 3$, it should be clear that Eqs. (5.24)-(5.26) determine the sol-gel coexistence region in the β-C plane for any value of f. For example, if $f = 5$, then Eq. (5.25) implies $\beta \geq 8/9$. Thus with higher valence antigens, cells with lower receptor densities can form cell surface receptor-ligand gels. Further analysis of Eqs. (5.24)-(5.26) shows that the sol-gel coexistence region becomes larger as f increases. Thus, for example, when $f = 3$ and $\beta = 4$, the gelation condition is met only at a single point, $KC^* = 1/3$. However, when $f = 5$, the gelation condition is met in a region, $2.14 \times 10^{-4} \leq KC \leq 6.42$.

CHAPTER 6

POST-GEL RELATIONS

<u>6.1</u> In this chapter we use the branching process approach to obtain the weight fraction distribution for finite-sized aggregates in systems containing a gel phase. Our approach is similar to that of Good (1963), Dobson and Gordon (1964), and Gordon and Malcolm (1966). It is based on the use of conditional probabilities. In particular, if E_1 and E_2 are two events, then the probability of E_1 conditional on E_2 is

$$P(E_1|E_2) = \frac{P(E_1 \cap E_2)}{P(E_2)} \quad . \tag{6.1}$$

By means of Eq. (6.1) we express conditional generating functions and probabilities in terms of their unconditional counterparts used until now, distinguishing conditional quantities by a circumflex (^).

<u>6.2</u> For a simple illustration of our method, we consider aggregation in systems of f-valent monomers. We first use Eq. (6.1) to calculate $\hat{w}_n$, the weight fraction of n-mers conditional on their being of finite size and therefore belonging to the sol. By analogy with Eq. (2.8), $\hat{w}_n$ is the probability that a rooted tree contains n nodes, conditional on the tree being finite, i.e.,

$$\hat{w}_n = P(Y = n | \text{tree is finite}) \quad . \tag{6.2}$$

The probability that a rooted tree is finite is simply $W(1) = F_0(u(1)) = W_s$ (see Sections 5.3 and 5.4). A tree which has n nodes is necessarily finite and hence by Eqs. (6.1) and (2.8)

$$\hat{w}_n = \frac{P(Y = n)}{W(1)} = \frac{w_n}{W(1)} \quad . \tag{6.3}$$

Thus $\hat{W}(\theta)$, the weight fraction generating function for molecules conditional on their belonging to the sol, is given by

$$\hat{W}(\theta) = \frac{W(\theta)}{W(1)} \quad . \tag{6.4}$$

When no gel is present, $W(1) = 1$ and $\hat{W}(\theta) = W(\theta)$.

6.3 For aggregates formed from antibodies and antigens, an analogous procedure may be followed to find $\hat{w}_{ij}^{(A)}$ (or $\hat{w}_{ij}^{(G)}$), the probability that a tree contains i antibodies and j antigens given that the tree is both finite and has an antibody (or antigen) as its root. To facilitate the use of conditional probabilities, let E_{ij} be the event that a rooted tree contains i antibodies and j antigens, E_A be the event that a tree has an antibody root, and E_F be the event that a rooted tree is finite. Then $\hat{w}_{ij}^{(A)} = P(E_{ij}|E_F \cap E_A)$, whereas $w_{ij}^{(A)} = P(E_{ij}|E_A)$. Also recall from Eq. (3.16) that $W_A(\underline{1}) = \sum_{i,j} w_{ij}^{(A)}$ and hence is the probability that a tree with an antibody root is finite. Symbolically, $W_A(\underline{1}) = P(E_F|E_A)$. Using Eq. (6.1) one finds

$$P(E_{ij}|E_F \cap E_A) = \frac{P(E_{ij} \cap E_F \ E_A)}{P(E_F \cap E_A)} = \frac{P(E_{ij} \cap E_F|E_A)P(E_A)}{P(E_F|E_A)P(E_A)} \quad .$$

A tree containing i antibodies and j antigens is necessarily finite. Therefore $P(E_{ij} \cap E_F|E_A) = P(E_{ij}|E_A)$. Hence

$$\hat{w}_{ij}^{(A)} = P(E_{ij}|E_F \cap E_A) = \frac{P(E_{ij}|E_A)}{P(E_F|E_A)} = \frac{w_{ij}^{(A)}}{W_A(\underline{1})} \quad . \tag{6.5a}$$

Similarly, one can show

$$\hat{w}_{ij}^{(G)} = \frac{w_{ij}^{(G)}}{W_G(\underline{1})} \quad . \tag{6.5b}$$

Thus, in terms of weight fraction generating functions

$$\hat{W}_A(\underline{\theta}) = W_A(\underline{\theta})/W_A(\underline{1}) \quad , \quad \hat{W}_G(\underline{\theta}) = W_G(\underline{\theta})/W_G(\underline{1}) \quad . \tag{6.6}$$

If division of one vector by another is defined elementwise [i.e., $\underline{x}/\underline{y} = (x_1/y_1, x_2/y_2, \ldots, x_n/y_n)$], then

$$\hat{\underline{W}}(\underline{\theta}) = \underline{W}(\underline{\theta})/\underline{W}(\underline{1}) \tag{6.7}$$

is the generalization of Eq. (6.4) to multitype processes.

6.4 Remark It is possible to develop an entire theory of branching processes conditional on trees being of finite size. In Appendix C we outline the procedures and derive various relationships between the conditional generating functions and their unconditional counterparts. Equation (6.7) is such a relationship.

6.5 Recognizing that W_s, the weight fraction of the sol, is the probability that a tree is finite, one finds in analogy with Eq. (3.15), that

$$W_s = \rho W_A(\underline{1}) + (1 - \rho)W_G(\underline{1}) \quad . \tag{6.8}$$

This equation can be derived by summing Eq. (3.15) over all values of i and j. In the pre-gel state $W_A(\underline{1}) = W_G(\underline{1}) = 1$, and hence $W_s = 1$ as one would anticipate.

6.6 To compute $\hat{w}_{ij}$, the weight fraction of aggregates with i antibodies and j antigens conditional on the aggregates belonging to the sol, $\hat{w}_{ij}^{(A)}$ and $\hat{w}_{ij}^{(G)}$ need to be combined. Because $\hat{w}_{ij}^{(A)}$ and $\hat{w}_{ij}^{(G)}$ are conditional on belonging to the sol, Eq. (3.15) needs to be modified to

$$\hat{w}_{ij} = \hat{\rho}\hat{w}_{ij}^{(A)} + (1 - \hat{\rho})\hat{w}_{ij}^{(G)} \quad , \tag{6.9}$$

where $\hat{\rho}$ is the probability that a randomly chosen root of a molecule in the sol is an antibody. For a system containing A antibodies and G antigens,

$$\hat{\rho} = \frac{W_A(\underline{1})A}{W_A(\underline{1})A + W_G(\underline{1})G} \quad . \tag{6.10}$$

Substituting Eqs. (6.5) and (6.10) into Eq. (6.9) and rewriting the result in terms of W_s [see Eq. (6.8)], one finds

$$\hat{w}_{ij} = \frac{w_{ij}(A + G)}{W_A(\underline{1})A + W_G(\underline{1})G} = \frac{w_{ij}}{W_s} \quad . \tag{6.11}$$

Thus the weight fraction distribution in the sol is simply a scaled version of the unconditional weight fraction distribution.

<u>6.7</u> **Remark** Equation (6.11) can be derived directly by the use of conditional probabilities, i.e.,

$$\hat{w}_{ij} = P(E_{ij}|E_F) = \frac{P(E_{ij} \cap E_F)}{P(E_F)} = \frac{P(E_{ij})}{P(E_F)} = \frac{w_{ij}}{W_s} \quad .$$

<u>6.8</u> To compute weight fractions in the sol using Eq. (6.11), we require $W_A(\underline{1})$ and $W_G(\underline{1})$. Recall from Eqs. (3.16)-(3.19)

$$\underline{W}(\underline{1}) = \underline{F}_0(\underline{v}) \tag{6.12}$$

where

$$\underline{v} \equiv \underline{u}(\underline{1}) = \underline{F}_1(\underline{v}) \quad . \tag{6.13}$$

Thus, $W_A(\underline{1}) = F_{A0}(\underline{v})$ and $W_G(\underline{1}) = F_{G0}(\underline{v})$, where $\underline{v} = (v_A, v_G)$. The components of $\underline{v}$ are the smallest positive roots of the equations

$$v_A = F_{A1}(v_A, v_G) \quad , \quad v_G = F_{G1}(v_A, v_G) \quad . \tag{6.14}$$

As shown in the next two examples and in Appendix C, determining $\underline{v}$ is the key to unraveling post-gel relations.

6.9 Example Antigen-Antibody Reactions with Equally Reactive Sites

In Section 3.11 we derived Goldberg's (1952) result for the weight fraction distribution of antigen-antibody aggregates in systems containing no gel. Here we extend that result, giving the condition for gelation and deriving weight-fractions in the sol. From Eqs. (6.14) and (3.24b) we find

$$v_A = 1-p_A + p_A v_G \quad , \quad v_G = (1-p_G + p_G v_A)^{f-1} \qquad (6.15)$$

and hence with Eq. (3.22)

$$v_G = (1-\alpha + \alpha v_G)^{f-1} \quad , \qquad (6.16)$$

where

$$\alpha \equiv p_A p_G = r p_G^2 \quad .$$

One solution to Eq. (6.16) is $v_G = 1$. Other solutions must be determined numerically when f is large. However, when $f = 2$ or 3, the additional solutions can easily be found algebraically.

First, let $f = 2$. Then

$$v_G = 1-\alpha + \alpha v_G \quad , \qquad (6.17)$$

and the only solution is $v_G = 1$. Substituting $v_G = 1$ into Eq. (6.15) gives $v_A = 1$, and hence $W_G(\underline{1}) = W_A(\underline{1}) = 1$, implying no gel is present. This same implication follows from the condition for gelation, Eq. (5.13). A gel phase exists when

$$\alpha > \alpha_c \equiv \frac{1}{f-1} \quad . \qquad (6.18)$$

Since $\alpha \leq 1$ always, Eq. (6.18) can never be satisfied when $f = 2$.

Second, let $f = 3$. Equation (6.16) has two solutions, namely, $v_G = 1$ and

$$v_G = \left(\frac{1-\alpha}{\alpha}\right)^2 \quad . \tag{6.19}$$

Substituting these two values of v_G in turn into Eq. (6.15) gives $v_A = 1$ and

$$v_A = 1 - p_A \frac{(2\alpha - 1)}{\alpha^2} \quad , \tag{6.20}$$

respectively. When $f = 3$, $\alpha > 1/2$ is the condition for gelation. As long as this condition is met, v_G and v_A, given by Eqs. (6.19) and (6.20), are less than one, and hence are the smallest positive roots of Eq. (6.14). Using Eqs. (6.19) and (6.20), the weight fraction of sol-phase aggregates with an antibody root becomes

$$W_A(\underline{1}) = F_{A0}(\underline{v}) = (1-p_A + p_A v_G)^2 = v_A^2 \quad ,$$

and the weight fraction of sol-phase aggregates with an antigen root becomes

$$W_G(\underline{1}) = F_{G0}(\underline{v}) = (1-p_G + p_G v_A)^3 = v_G^{3/2} \quad .$$

Thus for aggregates in the sol, Eq. (6.11) gives

$$\hat{w}_{ij} = \frac{w_{ij}(A + G)}{v_A^2 A + v_G^{3/2} G} \quad ,$$

where w_{ij} is given by Eq. (3.30).

6.10 Example Cell Surface Reactions

In the previous example, all sites were equally reactive. We now relax the assumption that sites act independently and utilize the model for multivalent antigens binding to and cross-linking cell surface receptors developed in Chapter 4, Section C. For this model, Eq. (4.2) requires

$$F_{A1}(v_A, v_G) = M_A^{-1} \sum_{k=1}^{2} k p_{Ak} v_G^{k-1} \quad ; \quad F_{G1}(v_A, v_G) = M_G^{-1} \sum_{i=1}^{f} i p_{Gi} v_A^{i-1} \quad . \tag{6.21}$$

Since the cell surface receptor sites are assumed to act independently, we have from Eqs. (4.20) and (4.21)

$$P_{Ak} = \binom{2}{k} p^k (1-p)^{2-k}$$

and

$$M_A = p_{A1} + 2p_{A2} = 2p \quad ,$$

where p, the probability that a receptor site is bound, is given by

$$p = \frac{S_0 - S}{S_0} \quad . \tag{6.22}$$

The generating function $F_{A1}(\underset{\sim}{v})$ thus reduces to

$$F_{A1}(v_G) = \frac{S}{S_0} + \left(1 - \frac{S}{S_0}\right) v_G \quad . \tag{6.23}$$

From Eqs. (4.27), (4.29), and (4.31)

$$p_{Gi} = \frac{KC}{fC_0} \binom{f}{i} \kappa^{i-1} S^i \quad , \quad i = 1, 2, \ldots, f \quad ,$$

and

$$M_G = (S_0 - S)/C_0 \quad .$$

Hence,

$$F_{G1}(v_A) = \frac{KCS}{f(S_0 - S)} \sum_{i=1}^{f} i \binom{f}{i} (\kappa S v_A)^{i-1}$$

$$= \frac{KCS}{S_0 - S} (1 + \kappa S v_A)^{f-1} \quad . \tag{6.24}$$

To determine the probabilities of extinction, v_A and v_G, we need to find the smallest positive roots of Eqs. (6.14). In the current example these equations become

$$v_A = \frac{S}{S_0} + \left(1 - \frac{S}{S_0}\right)v_G \quad , \tag{6.25}$$

and

$$v_G = \frac{KCS}{(S_0 - S)} (1 + \kappa S v_A)^{f-1} \quad . \tag{6.26}$$

Substituting Eq. (6.25) into Eq. (6.26) gives

$$v_G = \frac{KCS}{(S_0 - S)} \left[1 + \frac{\kappa S^2}{S_0} + \frac{\kappa S(S_0 - S)}{S_0} v_G \right]^{f-1} \quad . \tag{6.27}$$

Because S satisfies the conservation equation

$$S_0 = S + KCS(1 + \kappa S)^{f-1} \quad ,$$

Eq. (6.27) can be rewritten in the form

$$v_G = \left[\frac{S_0 + \kappa S^2 + \kappa S(S_0 - S)v_G}{S_0(1 + \kappa S)} \right]^{f-1} \quad . \tag{6.28}$$

When $f = 2$, the only solution to Eq. (6.28) is $v_G = 1$, and hence by Eq. (6.25), $v_A = 1$. When $f = 3$, the gelation criterion can be met for certain ranges of values of S (see Section 5.15). Further, in this case, Eq. (6.28) is quadratic in v_G, and hence can be solved analytically. Rather than solve for v_G directly, which yields an algebraically complicated expression, we take the square root of both sides of Eq. (6.28) and solve for

$$\gamma = v_G^{\frac{1}{2}} \quad . \tag{6.29}$$

Following this procedure, we find

$$S_0 + \kappa S^2 - S_0(1 + \kappa S)\gamma + \kappa S(S_0 - S)\gamma^2 = 0 \quad , \tag{6.30}$$

with solutions

$$\gamma = 1 \tag{6.31a}$$

and

$$\gamma = \frac{S_0 + \kappa S^2}{\kappa S(S_0 - S)} \quad . \tag{6.31b}$$

Because each solution of Eq. (6.30) must be a solution of Eq. (6.28), and because there are exactly two solutions of Eq. (6.28), the two values of γ, when squared, must be the values of v_G which solve Eq. (6.28).

Since $\gamma < 1$ when $v_G < 1$, it suffices to examine the conditions under which γ, given by Eq. (6.31b), is less than 1.

To find the range of values of S that correspond to $\gamma < 1$, we set $\gamma = 1$ in Eq. (6.31b) and solve for S. Not surprisingly, the two roots, which we shall call S^*, are precisely those that we predicted from the gelation criterion in Chapter 5 [see Eq. (5.28)]. In Fig. 6.1 we sketch γ as the function of S given by Eq. (6.31b). When $S \to 0$, $\gamma \cong 1/\kappa S \to \infty$, and when $S \to S_0$, $\gamma \cong S/(S_0 - S) \to \infty$. Hence the two values of S^* correspond to the crossings of the line $y = 1$ with the curve

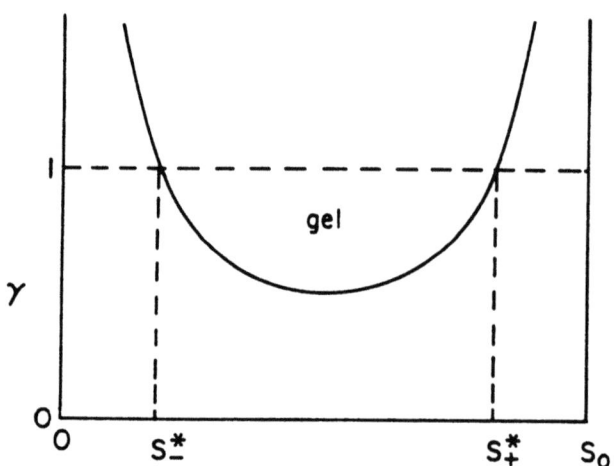

Figure 6.1

For values of S between S_-^* and S_+^*, $\gamma < 1$, and a gel phase exists.

$y = (S_0 + \kappa S^2)/(\kappa S(S_0 - S))$. For all S, $S_-^* < S < S_+^*$, $\gamma < 1$. The corresponding range of antigen concentrations for which a gel exists is given by Eq. (5.30).

In summary, when $f = 3$ and the gelation criteria $\kappa S_0 > 8$ [cf. Eq. (5.27)] and $C_-^* < C < C_+^*$ are met,

$$v_G = \left[\frac{S_0 + \kappa S^2}{\kappa S(S_0 - S)}\right]^2 < 1 \quad . \tag{6.32a}$$

Substituting v_G given by Eq. (6.32a) into Eq. (6.25) yields, after simplification,

$$v_A = \frac{S_0 + \kappa S^2(2 + \kappa S)}{\kappa^2 S^2} < 1 \quad . \tag{6.32b}$$

These values for v_A and v_G determine the weight fraction of sol-phase aggregates with a receptor root

$$W_A(\underline{1}) = F_{A0}(\underline{v}) = (1-p + pv_G)^2 = v_A^2 \quad , \tag{6.33}$$

and the weight fraction of sol-phase aggregates with an antigen root

$$W_G(\underline{1}) = F_{G0}(\underline{v}) = \sum_{i=0}^{3} p_{Gi} v_A^i$$

$$= \frac{C}{C_0} + \frac{KC}{3\kappa C_0} \sum_{i=1}^{3} \binom{3}{i}(\kappa S v_A)^i$$

$$= \frac{C}{C_0} + \frac{KC}{3\kappa C_0}\left[(1 + \kappa S v_A)^3 - 1\right] \quad , \tag{6.34}$$

where S is found by solving the conservation equation

$$S_0 = S + KCS(1 + \kappa S)^2 \quad .$$

Equation (6.11) now gives the weight fraction distribution of finite-sized (i,j)-mers on a cell surface in the presence of a gel.

CHAPTER 7

CONCLUSIONS AND EXTENSIONS

The branching process method as developed by Gordon (1962) and Good (1963) is an extremely useful and powerful technique for computing aggregate size distributions. In this paper we have explicated the method and its mathematical foundations in a manner which should make the technique accessible for application in biology. By way of example, we showed how the branching process method could be used to derive the weight fraction distribution of aggregates formed in the random polycondensation of f-valent particles. This distribution was originally obtained by Stockmayer (1943) using combinatorial and statistical mechanical techniques. As a further example, we derived the size distribution of antigen-antibody aggregates in solution. Contrasting the few lines we needed to obtain this result with the original computation by Goldberg (1952), who followed the Stockmayer procedure, shows the simplicity of the branching process method.

A valuable attribute of the branching process technique is the ability to solve for the distribution of aggregate sizes as a funcion of unspecified site occupancy probabilities. The solution to dynamical problems requires that these probabilities be given as functions of time, whereas for equilibrium calculations only the constant equilibrium values of these probabilities are required. Each choice of a set of site occupancy probabilities corresponds to a specific model of ligand binding and receptor cross-linking. For example, by choosing a binomial probability distribution for p_{Gk}, the probability that k ligand sites are occupied, one is implicitly assuming that the ligand sites do not interact and are filled independently of one another. Rather than pick site occupancy probability distributions in an _a priori_ manner, the method we have advocated is to build a kinetic (or equilibrium) model of ligand binding and receptor cross-linking incorporating any chemical assumptions about site interactions, steric hindrance, binding cooperativity, etc. From the model one then computes the site occupancy probabilities as a function of time or at equilibrium, as appropriate. This procedure was demonstrated in Chapter 4 using the binding model of Perelson and DeLisi (1980) for bivalent ligands.

For multivalent ligands, the binding model of Gandolfi, Giovenco, and Strom (1978), DeLisi (1980), and Perelson (1981) was used to obtain the major new result of this monograph: the weight fractions and concentrations of bivalent receptor-multivalent ligand aggregates as a function of the aggregate composition and biological parameters such as ligand concentration and valence, receptor affinity and density, and cross-linking equilibrium constants. For aggregates formed from multivalent ligands and g-valent receptors, $g \geq 2$, the appropriate generalizations are given in Appendix B. To derive receptor-ligand aggregate size distributions under an alternative binding and cross-linking model, one simply computes the relevant site occupancy probabilities from the model and substitutes them into the general expressions derived in Chapter 4.

Aggregates that form in idealized two-dimensional systems containing an infinite number of receptors can become arbitrarily large. The formation of such "infinite" complexes on an infinite two-dimensional membrane is the analogue of gelation in three-dimensional systems. In Chapter 5 we derived a criterion for two-dimensional gelation that we believe may be useful in characterizing the conditions under which macroscopic ligand-receptor aggregates form. As in three-dimensional systems, gelation occurs suddenly once a critical set of conditions are met. After the onset of gelation, the system can be thought of as being partitioned into two phases: a sol, containing only finite-sized aggregates, and a gel, comprising an infinite-sized aggregate.

Branching processes bring an interesting insight to the study of cell surface aggregation phenomena. The weight fraction distribution of finite-sized molecules changes as one goes from the pre-gel to the post-gel state. Gandolfi, Giovenco, and Strom (1978), DeLisi (1980), and Perelson (1981) neglected this change, which led to heretofore unexplained inconsistencies in their models. Based upon a simple procedure developed by Flory (cf. Flory, 1953), in Chapter 6 we used the branching process method to explicitly calculate in the post-gel state the weight fraction distribution of molecules belonging to the sol in three different situations: the random polycondensation of f-valent monomers, antigen-antibody reactions in solution, and cell surface cross-linking reactions. Because the Flory procedure treats ring formation within the gel in an approximate way, a more complex treatment would be required to obtain more accurate models of the post-gel state. The development of such models is an active area of research in physics and polymer chemistry. Various ideas involving percolation theory, critical exponents, and fractals are being examined for their ability to accurately describe the gel state (cf. Family and Landau, 1984).

The binding model employed in Chapter 4 was overly simplified as it did not allow intramolecular rearrangements of receptor-ligand aggregates. Intramolecular reactions generate rings that can increase the stability of an aggregate. Methods for addressing the general ring closure problem within the context of classical polymerization theory have been considered by Hoeve (1956), Gordon and Scantlebury (1966, 1967, 1968), Whittle (1965a,b, 1980a,b,c), Stepto (1982), Spouge (1984, 1985), and many others. In a separate publication we will show how to extend the branching process treatment given here to the situation in which an antibody (or receptor) binds at both of its binding sites to a single ligand. Recent experiments by Barisas (1984) indicate that such "monogamous bivalent attachments" of receptors to long, rigid, polymeric antigens may be commonplace. If a large fraction of receptor sites are bound in a monogamous bivalent manner, few receptor sites are available for cross-linking antigen molecules, and hence cluster sizes remain small.

Although our paper has been oriented toward cell surface aggregation phenomena, the branching process method is not restricted to problems in this area. Many other applications come to mind. For example, the aggregation in the presence of antigen of particles such as latex beads coated with antibody has been suggested as an immunoassay that could replace radioimmunoassays (cf. von Schulthess, et al., 1980). In order to quantitate such an assay, one requires a theory that relates the antigen concentration to the size of the resulting aggregate: the branching process method could provide such a theory. In related applications, microspheres or lipid vesicles coated with monoclonal antibodies and carrying pharmacologically active agents could be used to deliver lethal compounds or medicinal drugs to specific cells (cf. Rembaum and Dreyer, 1980). In order to deliver the appropriate amount of drug, one needs to predict, as a function of antibody concentration, the size of the resulting cell-microsphere clusters.

Yet another area in which the aggregation theory presented here may play a role is the analysis of cell-cell interactions. Many cells interact with other cells by means of specific molecules to form cellular aggregates. For example, red blood cells in the presence of certain macromolecules, such as fibrinogen, form long, cylindrical, and sometimes branched, aggregates called rouleaux. Rouleaux form in the human circulatory system and have been the object of intensive study. Application of the theory presented here can provide new predictions of the size distribution of rouleaux, possibly more accurate than have been obtained by other methods (cf. Perelson and Wiegel, 1982; Samsel and Perelson, 1982, 1984). In the immune system, the formation of cellular aggregates is an important primary step in the attack of foreign cells, virally infected cells, and tumor cells, by host defense cells such as T lymphocytes and macrophages (cf. Perelson and Macken, 1984). Quantitative measurements and mathematical theories of the size distribution of such aggregates are in their initial stages (cf. Segal and Stephany, 1984a,b; Perelson, 1985a). The branching process approach should prove of value in the analysis of these aggregates and in the more general problem of predicting the dynamics of formation and composition of any multicellular aggregate.

APPENDIX A

PROOF OF THEOREM 2

Assuming $F_{A1}(\underline{0}) \neq 0$ and $F_{G1}(\underline{0}) \neq 0$ we prove that Eq. (3.21a) holds. The proof of Eq. (3.21b) will then follow <u>mutatis mutandis</u>.

The general proof of this theorem does not apply if $i = 0$ or $j = 0$. Therefore, we deal with these cases first.

From the definition of $\underline{W}(\underline{\theta})$, for $i = 0$, $j \geq 1$

$$w_{0j}^{(A)} = C(\theta_G^j)\{W_A(\underline{\theta})\}$$

$$= 0 \quad , \quad j \geq 1 \quad . \tag{A.1}$$

However, when $i \geq 1$, $j = 0$, we have

$$w_{i0}^{(A)} = C(\theta_A^i)\{W_A(\underline{\theta})\}$$

$$= C(\theta_A^{i-1})\{F_{A0}(\underline{u}(\underline{\theta}))\}$$

$$= \begin{cases} p_{A0} & , \quad i = 1 \\ \left. \dfrac{\partial^{i-2}}{\partial \theta_A^{i-2}}\left(\dfrac{\partial F_{A0}(\underline{u})}{\partial u_A} \dfrac{\partial u_A}{\partial \theta_A}\right)\right|_{\theta_A = 0} & , \quad i \geq 2 \end{cases} \quad .$$

Because particle types alternate with generation in antibody-antigen reactions, F_{A0} is a function of u_G only, and so $\partial F_{A0}(\underline{u})/\partial u_A = 0$. Hence

$$w_{i0}^{(A)} = \begin{cases} p_{A0} & , \quad i = 1 \\ 0 & , \quad i \geq 2 \end{cases} \quad . \tag{A.2}$$

It remains to prove for $i \geq 1$ and $j \geq 1$,

$$C(\theta_A^{i-1}\theta_G^j)\{F_{A0}(\underset{\sim}{u}(\underset{\sim}{\theta}))\} = \begin{cases} \dfrac{M_A}{(i-1)j} C(u_A^{i-2}) \left\{\dfrac{d[F_{G1}^j(\underset{\sim}{u})]}{du_A}\right\} C(u_G^{j-1})\{F_{A1}^i(\underset{\sim}{u})\}, & i \geq 2, j \geq 1 \\ \\ C(u_A^0)\{F_{G1}^j(\underset{\sim}{u})\} C(u_G^j)\{F_{A0}(\underset{\sim}{u})\} & , i = 1, j \geq 1, \end{cases}$$

where

$$F_{A0}(\underset{\sim}{u}) = \sum_{k=0}^{2} p_{Ak} u_G^k \quad , \quad F_{G0}(\underset{\sim}{u}) = \sum_{k=0}^{f} p_{Gk} u_A^k \quad , \tag{A.3}$$

$$F_{A1}(\underset{\sim}{u}) = \sum_{k=0}^{1} \tilde{p}_{Ak} u_G^k \quad , \quad F_{G1}(\underset{\sim}{u}) = \sum_{k=0}^{f-1} \tilde{p}_{Gk} u_A^k \quad , \tag{A.4}$$

$$M_A = \left.\frac{\partial F_{A0}(\underset{\sim}{u})}{\partial u_G}\right|_{\underset{\sim}{u}=\underset{\sim}{1}} \quad , \quad M_G = \left.\frac{\partial F_{G0}(\underset{\sim}{u})}{\partial u_A}\right|_{\underset{\sim}{u}=\underset{\sim}{1}} \quad . \tag{A.5}$$

Our proof starts with a result due to Good (1960, Theorem 8; 1965, Theorem A) for multitype branching processes which we state in our notation as it applies to a two-type system.

<u>Theorem (Good)</u> Suppose

$$\theta_A = \frac{u_A}{F_{A1}(\underset{\sim}{u})} \quad , \quad \theta_G = \frac{u_G}{F_{G1}(\underset{\sim}{u})} \tag{A.6}$$

where $F_{A1}(\underset{\sim}{u})$ and $F_{G1}(\underset{\sim}{u})$ are analytic in a neighborhood of the origin, $F_{A1}(\underset{\sim}{0}) \neq 0$ and $F_{G1}(\underset{\sim}{0}) \neq 0$. Further let $G_{A0}(\underset{\sim}{u},\underset{\sim}{\theta})$ be a function which is analytic in a neighborhood of $(\underset{\sim}{u},\underset{\sim}{\theta}) = (\underset{\sim}{0},\underset{\sim}{0})$ except at poles; i.e., let $G_{A0}(\underset{\sim}{u},\underset{\sim}{\theta})$ be meromorphic. Then

$$C(\theta_A^i \theta_G^j)\{G_{A0}(\underset{\sim}{u}(\underset{\sim}{\theta}),\underset{\sim}{\theta})\} = C(u_A^i u_G^j)\{G_{A0}(\underset{\sim}{u},\underset{\sim}{\theta}(\underset{\sim}{u}))F_{A1}^i(\underset{\sim}{u})F_{G1}^j(\underset{\sim}{u})\det J\} \quad , \tag{A.7}$$

where det J is the determinant of the 2 × 2 matrix

$$J = \begin{bmatrix} 1 & \dfrac{-u_A}{F_{A1}(\underline{u})} \dfrac{\partial F_{A1}(\underline{u})}{\partial u_G} \\ \dfrac{-u_G}{F_{G1}(\underline{u})} \dfrac{\partial F_{G1}(\underline{u})}{\partial u_A} & 1 \end{bmatrix}$$

We apply the theorem with $G_{A0}(\underline{u},\underline{\theta}) = F_{A0}(\underline{u}(\underline{\theta}))$. Observe that the analyticity of $F_{A1}(\underline{u})$ and $F_{G1}(\underline{u})$ in a neighborhood of $\underline{u} = \underline{0}$ is clear from their definitions as probability generating functions. In addition, $F_{A1}(\underline{0}) = \tilde{p}_{A0} \neq 0$ and $F_{G1}(\underline{0}) = \tilde{p}_{G0} \neq 0$ by assumption. To show that $G_{A0}(\underline{u},\underline{\theta})$ is analytic in a neighborhood of the origin we write, using Eqs. (A.3) and (A.6),

$$G_{A0}(\underline{u},\underline{\theta}) = F_{A0}(\underline{u}(\underline{\theta})) = \sum_{k=0}^{2} p_{Ak} \theta_G^k F_{G1}^k(\underline{u})\;.$$

First considered as a function of $\underline{\theta}$, $G_{A0}(\underline{u},\underline{\theta})$ is a convergent power series and hence analytic in a neighborhood of $\underline{\theta} = \underline{0}$. Next considered as a function of $\underline{u}$, $G_{A0}(\underline{u},\underline{\theta})$ has the properties of $F_{G1}(\underline{u})$. Since $F_{G1}(\underline{u})$ is analytic in a neighborhood of $\underline{u} = \underline{0}$, $G_{A0}(\underline{u},\underline{\theta})$ has this property.

Now, from the definition of $\underline{W}(\underline{\theta})$, for $i \geq 1$, $j \geq 1$, and Eq. (A.7)

$$\begin{aligned} w_{ij}^{(A)} &= C(\theta_A^{i-1}\theta_G^j)\{F_{A0}(\underline{u}(\underline{\theta}))\} \\ &= C(u_A^{i-1}u_G^j)\{F_{A0}(\underline{u})F_{A1}^{i-1}(\underline{u})F_{G1}^j(\underline{u})\det J\}\;. \end{aligned} \quad (A.8)$$

We first evaluate $\det J$. Because particle types alternate with generation in antibody-antigen reactions, F_{A0} and F_{A1} are functions of u_G only, and F_{G0} and F_{G1} are functions of u_A only; i.e.,

$$\partial F_{Aj}(\underline{u})/\partial u_A = 0\;, \quad \partial F_{Gj}(\underline{u})/\partial u_G = 0\;, \quad j = 0,1\;. \quad (A.9)$$

Hence

$$\det J = 1 - \frac{u_A u_G}{F_{A1} F_{G1}} \frac{dF_{A1}}{du_G} \frac{dF_{G1}}{du_A}\;. \quad (A.10)$$

We distinguish the cases $i = 1$ and $i > 1$ where $j \geq 1$ in both cases. For $i > 1$, substitute det J into Eq. (A.8). Write $F_{A1}^{i-2}(dF_{A1}/du_G)$ as $(i-1)^{-1}(d[F_{A1}^{i-1}]/du_G)$ and $F_{G1}^{j-1}(dF_{G1}/du_A)$ as $j^{-1}(d[F_{G1}^j]/du_A)$ to obtain

$$w_{ij}^{(A)} = C(u_A^{i-1}u_G^j)\left\{F_{A0}F_{A1}^{i-1}F_{G1}^j - \frac{u_A u_G}{(i-1)j}F_{A0}\frac{d[F_{A1}^{i-1}]}{du_G}\frac{d[F_{G1}^j]}{du_A}\right\}$$

$$= \frac{1}{(i-1)!j!}\left[\frac{\partial^{i+j-3}}{\partial u_A^{i-2}\partial u_G^{j-1}}\left\{\frac{\partial^2}{\partial u_A \partial u_G}\left[F_{A0}F_{A1}^{i-1}F_{G1}^j\right]\right\}\right]_{\underline{u}=\underline{0}}$$

$$- \frac{1}{(i-1)j}C(u_A^{i-2}u_G^{j-1})\left\{F_{A0}\frac{d[F_{A1}^{i-1}]}{du_G}\frac{d[F_{G1}^j]}{du_A}\right\} \quad . \tag{A.11}$$

Appealing again to Eq. (A.9), we reduce (A.11) to

$$w_{ij}^{(A)} = \frac{1}{(i-1)!j!}\left[\frac{\partial^{i+j-3}}{\partial u_A^{i-2}\partial u_G^{j-1}}\left(\frac{dF_{A0}}{du_G}F_{A1}^{i-1}\frac{d[F_{G1}^j]}{du_A} + F_{A0}\frac{d[F_{A1}^{i-1}]}{du_G}\frac{d[F_{G1}^j]}{du_A}\right)\right]_{\underline{u}=\underline{0}}$$

$$- \frac{1}{(i-1)j}C(u_A^{i-2}u_G^{j-1})\left\{F_{A0}\frac{d[F_{A1}^{i-1}]}{du_G}\frac{d[F_{G1}^j]}{du_A}\right\}$$

$$= \frac{1}{(i-1)j}C(u_A^{i-2}u_G^{j-1})\left\{\frac{dF_{A0}}{du_G}F_{A1}^{i-1}\frac{d[F_{G1}^j]}{du_A} + F_{A0}\frac{d[F_{A1}^{i-1}]}{du_G}\frac{d[F_{G1}^j]}{du_A}\right\}$$

$$- \frac{1}{(i-1)j}C(u_A^{i-2}u_G^{j-1})\left\{F_{A0}\frac{d[F_{A1}^{i-1}]}{du_G}\frac{d[F_{G1}^j]}{du_A}\right\} \quad .$$

That is,

$$w_{ij}^{(A)} = \frac{1}{(i-1)j}C(u_A^{i-2}u_G^{j-1})\left\{\frac{dF_{A0}}{du_G}F_{A1}^{i-1}\frac{d[F_{G1}^j]}{du_A}\right\} \quad . \tag{A.12}$$

From the universal consistency relation, Eq. (3.9), and the definition of M_A, Eq. (3.11),

$$\frac{dF_{A0}}{du_G} = M_A F_{A1}(u_G) \quad.$$

Thus

$$w_{ij}^{(A)} = \frac{1}{(i-1)j} C(u_A^{i-2}) \left\{ \frac{d[F_{G1}^j]}{du_A} \right\} C(u_G^{j-1})\{M_A F_{A1}^i\} \quad, \quad i \geq 2 \quad, \quad j \geq 1 \quad. \quad (A.13)$$

Now, for $i = 1$, Eqs. (A.8) and (A.10) imply

$$w_{1j}^{(A)} = C(u_A^0 u_G^j) \left\{ F_{A0}(\underline{u}) \; F_{G1}^j(\underline{u}) \left(1 - \frac{u_A u_G}{F_{A1}(\underline{u})F_{G1}(\underline{u})} \frac{dF_{A1}(\underline{u})}{du_G} \frac{dF_{G1}(\underline{u})}{du_A} \right) \right\}$$

$$= C(u_A^0 u_G^j)\{F_{A0}(\underline{u}) \; F_{G1}^j(\underline{u})\}$$

$$= C(u_A^0)\{F_{G1}^j(\underline{u})\}C(u_G^j)\{F_{A0}(\underline{u})\} \quad. \quad (A.14)$$

Finally, notice that Eq. (A.14) is also valid when $j = 0$ since it reduces to the particular result of Eq. (A.2).

APPENDIX B

INTERACTIONS OF g-VALENT ANTIBODY WITH f-VALENT ANTIGEN

We apply Theorem 2 (Section 3.9) to find weight fractions of aggregates of g-valent antibodies and f-valent antigens, f, g $\geq$ 1. Starting with the most general situation where neither antibody nor antigen sites are assumed to be equally reactive, we later specialize to the case where antibody sites are independent, and finally to the case where both antibody and antigen sites are independent.

Let

$$p_{Ak} = P(k \text{ sites bound on an antibody root}), \quad k = 0, 1, \ldots, g \quad ,$$

and

$$p_{Gk} = P(k \text{ sites bound on an antigen root}), \quad k = 0, 1, \ldots, f \quad .$$

Then

$$F_{A0}(\underline{\theta}) = \sum_{k=0}^{g} p_{Ak} \theta_G^k \quad , \quad F_{G0}(\underline{\theta}) = \sum_{k=0}^{f} p_{Gk} \theta_A^k \quad ; \tag{B.1}$$

$$F_{A1}(\underline{\theta}) = \sum_{k=0}^{g-1} \tilde{p}_{Ak} \theta_G^k \quad , \quad F_{G1}(\underline{\theta}) = \sum_{k=0}^{f-1} \tilde{p}_{Gk} \theta_A^k \quad ; \tag{B.2}$$

where

$$\tilde{p}_{Ak} = M_A^{-1}(k+1)p_{Ak} \quad , \quad \tilde{p}_{Gk} = M_G^{-1}(k+1)p_{Gk} \tag{B.3}$$

and

$$M_A = \sum_{k=1}^{g} k p_{Ak} \quad , \quad M_G = \sum_{k=1}^{f} k p_{Gk} \quad . \tag{B.4}$$

To simplify the expressions for $w_{ij}^{(A)}$ and $w_{ij}^{(G)}$, we shall write

$$P_{ab}^{(A)} = \sum_{\underset{\sim}{x}} \binom{a}{\underset{\sim}{x}} \prod_{\ell=1}^{g} \left(M_A^{-1} \ell p_{A\ell}\right)^{x_\ell} \tag{B.5}$$

where $\underset{\sim}{x} = (x_1, x_2, \ldots, x_g)$ satisfies the conditions

$$x_\ell \geq 0 \quad , \quad \ell = 1, 2, \ldots, g \quad ,$$

$$\sum_{\ell=1}^{g} x_\ell = a \quad ,$$

$$\sum_{\ell=1}^{g} (\ell-1) x_\ell = b \quad .$$

The right side of Eq. (B.5) gives the probability that a antibody "parents" all in generation r, $r \geq 1$, bear b antigen "offspring." There is an analogous probability for antigen "parents," namely

$$P_{ab}^{(G)} = \sum_{\underset{\sim}{x}} \binom{a}{\underset{\sim}{x}} \prod_{\ell=1}^{f} \left(M_G^{-1} \ell p_{G\ell}\right)^{x_\ell} \tag{B.6}$$

where $\underset{\sim}{x} = (x_1, x_2, \ldots, x_f)$ satisfies the conditions

$$x_\ell \geq 0 \quad , \quad \ell = 1, 2, \ldots, f \quad ,$$

$$\sum_{\ell=1}^{f} x_\ell = a \quad ,$$

$$\sum_{\ell=1}^{f} (\ell-1) x_\ell = b \quad .$$

Then

$$w_{ij}^{(A)} = \begin{cases} \dfrac{M_A}{(i-1)} P_{i,j-1}^{(A)} \sum_{k=2}^{\min(i,f)} (k-1)\tilde{p}_{G,k-1} P_{j-1,i-k}^{(G)} , & i \geq 2 , j \geq 1 \\ 0 & , i \geq 2 , j = 0 \\ p_{Aj}(M_G^{-1} p_{G1})^j & , i = 1 , j \geq 0 \\ 0 & , i = 0 , j \geq 0 \end{cases} \quad (B.7a)$$

and

$$w_{ij}^{(G)} = \begin{cases} \dfrac{M_G}{(j-1)} P_{j,i-1}^{(G)} \sum_{k=2}^{\min(j,g)} (k-1)\tilde{p}_{A,k-1} P_{i-1,j-k}^{(A)} , & j \geq 2 , i \geq 1 \\ 0 & , j \geq 2 , i = 0 \\ p_{Gi}(M_A^{-1} p_{A1})^i & , j = 1 , i \geq 0 \\ 0 & , j = 0 , i \geq 0 \end{cases} \quad (B.7b)$$

Antibody Sites Equally Reactive

If antibody sites are equally reactive, with p_A the probability of a site being bound, then

$$p_{Aj} = \binom{g}{j}(1-p_A)^{g-j} p_A^j , \quad j = 0, 1, \ldots, g , \quad (B.8)$$

$$M_A = g p_A , \quad (B.9)$$

and

$$P_{ab}^{(A)} = \binom{(g-1)a}{b}(1-p_A)^{(g-1)a-b} p_A^b . \quad (B.10)$$

Equations (B.8)-(B.10) are substituted into Eq. (B.7). The form for $w_{ij}^{(G)}$ when $j \geq 2$ and $i \geq 1$, can then be shown by direct substitution to give the correct expression for $w_{ij}^{(G)}$ when $j = 1$ and $i \geq 1$. Thus, these cases are combined below. After some simplification, we obtain

$$w_{ij}^{(A)} = \begin{cases} \dfrac{g}{(i-1)} \binom{(g-1)i}{j-1} (1-p_A)^{(g-1)i-j+1} p_A^j \sum_{k=2}^{\min(i,f)} (k-1)\tilde{p}_{G,k-1} P_{j-1,i-k}^{(G)} \,, \\ \qquad\qquad\qquad\qquad\qquad\qquad\qquad\qquad i \geq 2 \,, \; j \geq 1 \\ 0 \qquad\qquad\qquad\qquad\qquad\qquad\qquad\quad\,, \; i \geq 2 \,, \; j = 0 \\ \binom{g}{j} (1-p_A)^{g-j} p_A^j (M_G^{-1} p_{G1})^j \qquad\,, \; i = 1 \,, \; j \geq 0 \\ 0 \qquad\qquad\qquad\qquad\qquad\qquad\qquad\quad\,, \; i = 0 \,, \; j \geq 0 \end{cases} \quad \text{(B.11a)}$$

and

$$w_{ij}^{(G)} = \begin{cases} \dfrac{M_G}{i} \binom{(g-1)i}{j-1} (1-p_A)^{(g-1)i-j+1} p_A^{j-1} P_{j,i-1}^{(G)} \,, \\ \qquad\qquad\qquad\qquad\qquad\qquad j \geq 1 \,, \; i \geq 1 \\ 0 \qquad\qquad\qquad\qquad\qquad\,, \; j \geq 2 \,, \; i = 0 \\ p_{G0} \qquad\qquad\qquad\qquad\quad\,\,, \; j = 1 \,, \; i = 0 \\ 0 \qquad\qquad\qquad\qquad\qquad\,, \; j = 0 \,, \; i \geq 0 \end{cases} \quad \text{(B.11b)}$$

Antibody and Antigen Sites Equally Reactive

If, in addition, antigen sites are equally reactive with p_G, the probability of a site being bound, then

$$p_{Gi} = \binom{f}{i} (1-p_G)^{f-i} p_G^i \,, \quad i = 0, 1, \ldots, f \,, \tag{B.12}$$

$$M_G = f p_G \,, \tag{B.13}$$

and

$$P_{ab}^{(G)} = \binom{(f-1)a}{b} (1-p_G)^{(f-1)a-b} p_G^b \,. \tag{B.14}$$

Equations (B.12)-(B.14) are substituted into Eq. (B.11). Again, the form for $w_{ij}^{(A)}$, when $i \geq 2$ and $j \geq 1$, is also correct for $i \geq 1$. Thus, we combine cases to obtain

$$w_{ij}^{(A)} = \begin{cases} \dfrac{g}{j}\dbinom{(g-1)i}{j-1}\dbinom{(f-1)j}{i-1}(1-p_A)^{(g-1)i-j+1}p_A^j(1-p_G)^{(f-1)j-i+1}p_G^{i-1} \\ \qquad\qquad\qquad\qquad\qquad\qquad\qquad\qquad\qquad\qquad , \quad i \geq 1 \;,\; j \geq 1 \\ 0 \qquad\qquad\qquad\qquad\qquad\qquad\qquad\qquad , \quad i \geq 2 \;,\; j = 0 \\ p_{A0} \qquad\qquad\qquad\qquad\qquad\qquad\qquad , \quad i = 1 \;,\; j = 0 \\ 0 \qquad\qquad\qquad\qquad\qquad\qquad\qquad\qquad , \quad i = 0 \;,\; j \geq 0 \end{cases} \quad (B.15a)$$

and

$$w_{ij}^{(G)} = \begin{cases} \dfrac{f}{i}\dbinom{(g-1)i}{j-1}\dbinom{(f-1)j}{i-1}(1-p_A)^{(g-1)i-j+1}p_A^{j-1}(1-p_G)^{(f-1)j-i+1}p_G^i \\ \qquad\qquad\qquad\qquad\qquad\qquad\qquad\qquad\qquad\qquad , \quad j \geq 1 \;,\; i \geq 1 \\ 0 \qquad\qquad\qquad\qquad\qquad\qquad\qquad\qquad , \quad j \geq 2 \;,\; i = 0 \\ p_{G0} \qquad\qquad\qquad\qquad\qquad\qquad\qquad , \quad j = 1 \;,\; i = 0 \\ 0 \qquad\qquad\qquad\qquad\qquad\qquad\qquad\qquad , \quad j = 0 \;,\; i \geq 0 \end{cases} \quad (B.15b)$$

Combining $w_{ij}^{(A)}$ and $w_{ij}^{(G)}$ according to Eq. (3.15), choosing $\rho = A/(A+G)$ as in Eq. (3.29), and using the substitution $p_A = rp_G$ where $r = fG/gA$, gives w_{ij}.

Thus

$$w_{ij} = \rho w_{ij}^{(A)} + (1-\rho)w_{ij}^{(G)} \quad,$$

and hence

$$w_{0j} = \rho w_{0j}^{(A)} + (1-\rho)w_{0j}^{(G)}$$

$$= \begin{cases} \dfrac{G}{A+G}(1-p_G)^f \;, & j = 1 \;, \\ 0 \;, & j \geq 2 \;, \end{cases} \quad (B.16a)$$

$$w_{i0} = \rho w_{i0}^{(A)}$$

$$= \begin{cases} \dfrac{A}{A+G}(1-p_A)^g \;, & i = 1 \;, \\ 0 \;, & i \geq 2 \;, \end{cases} \quad (B.16b)$$

and

$$w_{ij} = \rho w_{ij}^{(A)} + (1-\rho)w_{ij}^{(G)}$$

$$= \frac{fG(i+j)}{(A+G)ij}\binom{(g-1)i}{j-1}\binom{(f-1)j}{i-1}(1-rp_G)^{(g-1)i-j+1}r^{j-1} \cdot$$

$$(1-p_G)^{(f-1)j-i+1}p_G^{i+j-1} \quad , \quad i \geq 1 \quad , \quad j \geq 1 \quad . \quad (B.16c)$$

If this set of weight fractions is converted into the concentrations m_{ij} via Eq. (3.31), one finds that for all i and j, with $i + j \neq 0$,

$$m_{ij} = \frac{fG[(g-1)i]![(f-1)j]!(1-rp_G)^{(g-1)i-j+1}r^{j-1}(1-p_G)^{(f-1)j-i+1}p_G^{i+j-1}}{j![(g-1)i - j + 1]!i![(f-1)j - i + 1]!}$$

(B.17)

which agrees with the result of Goldberg (1953).

Antigen But Not Antibody Sites Equally Reactive

For completeness, we give formulae for the situation where antigen, but not antibody, sites are equally reactive. To this end, substitute Eqs. (B.12)-(B.14) into Eq. (B.7) and obtain

$$w_{ij}^{(A)} = \begin{cases} \dfrac{M_A}{j}\binom{(f-1)j}{i-1}(1-p_G)^{(f-1)j-i+1}p_G^{i-1}\,P_{i,j-1}^{(A)} \;, & i \geq 1 \;,\; j \geq 1 \\ 0 & ,\; i \geq 2 \;,\; j = 0 \\ P_{A0} & ,\; i = 1 \;,\; j = 0 \\ 0 & ,\; i = 0 \;,\; j \geq 0 \end{cases} \quad (B.18a)$$

and

$$w_{ij}^{(G)} = \begin{cases} \dfrac{f}{j-1} \binom{(f-1)j}{i-1} (1-p_G)^{(f-1)j-i+1} p_G^i \sum_{k=2}^{\min(j,g)} (k-1)\tilde{p}_{A,k-1} P_{i-1,j-k}^{(A)} &, \quad j \geq 2 \;,\; i \geq 1 \\ 0 &, \quad j \geq 2 \;,\; i = 0 \\ \binom{f}{i} (1-p_G)^{f-1} p_G^i (M_A^{-1} p_{A1})^i &, \quad j = 1 \;,\; i \geq 0 \\ 0 &, \quad j = 0 \;,\; i \geq 0 \end{cases} \quad \text{(B.18b)}$$

APPENDIX C

GENERATING FUNCTIONS FOR POST-GEL RELATIONS

Here we derive generating functions appropriate to finite-sized antibody-antigen aggregates belonging to the sol phase of a post-gel system, and illustrate their use by formally deriving Eq. (6.7).

Fundamental to these conditional generating functions are the following definitions of conditional offspring probability distributions. Let $\hat{p}_{Ak}$ be the probability that an antibody in generation zero has k offspring, conditional on the aggregate of which the antibody is a member belonging to the sol. Then by Eq. (6.1)

$$\hat{p}_{Ak} = \frac{p_{Ak} v_G^k}{\sum_{n=0}^{2} p_{An} v_G^n} \quad , \tag{C.1}$$

where v_G represents the probability that an antigen in the first or a later generation has a finite number of descendents. To understand Eq. (C.1), observe that the term p_{Ak} in the numerator is simply the unconditional probability that an antibody in generation zero has k offspring. Each of these offspring is an antigen which, with probability v_G, has a finite number of descendents. The denominator correctly normalizes the equation. In an analogous manner we define $\hat{p}_{G\ell}$, the probability that an antigen in generation zero has ℓ offspring, conditional on the aggregate of which the antigen is a member belonging to the sol. Thus

$$\hat{p}_{G\ell} = \frac{p_{G\ell} v_A^\ell}{\sum_{n=0}^{f} p_{Gn} v_A^n} \quad , \tag{C.2}$$

where v_A represents the probability that an antibody in the first or later generation has a finite number of descendents.

Clearly the probability vector $\underline{v} = (v_A, v_G)$ is the key to post-gel relations. We determine $\underline{v}$ by solving

$$\underline{u}(\underline{1}) \equiv \underline{v} = \underline{F}_1(\underline{v}) \ . \tag{C.3}$$

The solutions which are relevant to gelation are the lowest positive roots of the equations

$$v_A = F_{A1}(v_A, v_G) \ , \quad v_G = F_{G1}(v_A, v_G) \ . \tag{C.4}$$

Based on $\hat{p}_{Ak}$ and $\hat{p}_{G\ell}$ the methods of Chapters 3 and 4 can be duplicated to derive various relations applicable to aggregates in the sol. To begin, we introduce $\hat{\underline{F}}_r(\underline{\theta})$, the generating function for the number of offspring contributed by a particle in generation r, conditional on the aggregate to which the particle belongs being finite. By analogy with Eqs. (3.7)-(3.10) let

$$\hat{F}_{A0}(\underline{\theta}) = \sum_{k=0}^{2} \hat{p}_{Ak} \theta_G^k \ , \quad \hat{F}_{G0}(\underline{\theta}) = \sum_{\ell=0}^{f} \hat{p}_{G\ell} \theta_A^\ell \ , \tag{C.5a}$$

and

$$\hat{F}_{A1}(\underline{\theta}) = \hat{M}_A^{-1} \sum_{k=1}^{2} k \hat{p}_{Ak} \theta_G^{k-1} \ , \quad \hat{F}_{G1}(\underline{\theta}) = \hat{M}_G^{-1} \sum_{\ell=1}^{f} \ell \hat{p}_{G\ell} \theta_A^{\ell-1} \ , \tag{C.5b}$$

where

$$\hat{M}_A = \hat{F}'_{A0}(\underline{1}) \quad \text{and} \quad \hat{M}_G = \hat{F}'_{G0}(\underline{1}) \tag{C.5c}$$

are the respective mean number of offspring from an antibody and antigen in generation zero, conditional on the aggregate being in the sol. By analogy with Eqs. (3.16)-(3.19), we construct the weight fraction generating function for aggregates in the sol

$$\hat{\underline{W}}(\underline{\theta}) = \underline{\theta} \otimes \hat{\underline{F}}_0(\hat{\underline{u}}(\underline{\theta})) \ , \tag{C.6}$$

where

$$\hat{\underline{u}}(\underline{\theta}) = \underline{\theta} \otimes \hat{\underline{F}}_1(\hat{\underline{u}}(\underline{\theta})) \quad . \tag{C.7}$$

Finally, Theorem 2 (Section 3.9) can be used to express the coefficients $\hat{w}_{ij}^{(A)}$ and $\hat{w}_{ij}^{(G)}$, and hence the weight fraction distribution, in terms of the unconditional offspring distributions, p_{Ak} and $p_{G\ell}$. These unconditional probability distributions are either empirically measured or modeled in terms of measurable quantities (see Chapter 4) so that, with perhaps some mathematical difficulty, conditional weight fraction distributions may be computed. However, if one knows $w_{ij}^{(A)}$ and $w_{ij}^{(G)}$, the weight fractions not conditioned on being in the sol, a simpler, more direct procedure can be followed as shown in Section 6.6.

An alternative to direct application of Theorem 2 to Eqs. (C.5)-(C.7) is afforded by noting the following simplifying relations which relate $\hat{\underline{F}}_0(\underline{\theta})$ and $\hat{\underline{F}}_1(\underline{\theta})$ to their unconditional counterparts, $\underline{F}_0(\underline{\theta})$ and $\underline{F}_1(\underline{\theta})$. For example, substituting Eq. (C.1) into Eq. (C.5a) leads to

$$\hat{F}_{A0}(\underline{\theta}) = \frac{\sum_{k=0}^{2} p_{Ak} v_G^k \theta_G^k}{\sum_{k=0}^{2} p_{Ak} v_G^k} = F_{A0}(\underline{v} \otimes \underline{\theta})/F_{A0}(\underline{v}) \quad . \tag{C.8}$$

A similar relation can be derived for $\hat{F}_{G0}(\underline{\theta})$. Hence, as Gordon and Malcolm (1966) noted,

$$\hat{\underline{F}}_0(\underline{\theta}) = \underline{F}_0(\underline{v} \otimes \underline{\theta})/\underline{F}_0(\underline{v}) \quad , \tag{C.9}$$

where division of two n-vectors $\underline{x}$ and $\underline{y}$ is defined by

$$\underline{x}/\underline{y} = (x_1/y_1, x_2/y_2, \ldots, x_n/y_n) \quad . \tag{C.10}$$

Analogously, one can show

$$\hat{F}_{A1}(\underline{\theta}) = \sum_{k=1}^{2} k p_{Ak} (v_G \theta_G)^{k-1} / \sum_{k=1}^{2} k p_{Ak} v_G^{k-1} = F'_{A0}(\underline{v} \otimes \underline{\theta})/F'_{A0}(\underline{v}) \quad , \tag{C.11a}$$

$$\hat{F}_{G1}(\underline{\theta}) = F'_{G0}(\underline{v} \otimes \underline{\theta})/F'_{G0}(\underline{v}) \quad , \tag{C.11b}$$

and thus,

$$\hat{\underline{F}}_1(\underline{\theta}) = \underline{F}'_0(\underline{v} \otimes \underline{\theta})/\underline{F}'_0(\underline{v}) \quad , \tag{C.11c}$$

where in using the prime notation for derivatives we are cognizant of the fact that $F_{A0}(\underline{\theta}) = F_{A0}(\theta_G)$ and $F_{G0}(\underline{\theta}) = F_{G0}(\theta_A)$.

Another simplifying relation between conditional and unconditional generating functions is

$$\hat{\underline{W}}(\underline{\theta}) = \underline{W}(\underline{\theta})/\underline{W}(\underline{1}) \quad , \tag{C.12}$$

derived in Section 6.3 by formal probability arguments. Here we derive the same result starting with the generating functions above. From Eqs. (C.6) and (C.9)

$$\hat{\underline{W}}(\underline{\theta}) = \underline{\theta} \otimes \underline{F}_0(\underline{v} \otimes \hat{\underline{u}}(\underline{\theta}))/\underline{F}_0(\underline{v}) \quad .$$

By Eq. (6.12), $\underline{F}_0(\underline{v}) = \underline{W}(\underline{1})$ and by Eq. (3.18), $\underline{W}(\underline{\theta}) = \underline{\theta} \otimes \underline{F}_0(\underline{u}(\underline{\theta}))$. Thus if we can show

$$\underline{v} \otimes \hat{\underline{u}}(\underline{\theta}) = \underline{u}(\underline{\theta}) \quad ,$$

then the derivation is complete. We shall need the relations

$$\underline{v} = \underline{F}_1(\underline{v}) = \underline{F}'_0(\underline{v})/\underline{F}'_0(\underline{1}) \tag{C.13}$$

that follow from Eq. (6.13) and the universal consistency relations, Eq. (3.9). Now, by Eqs. (C.7) and (C.11)

$$\hat{\underline{u}}(\underline{\theta}) = \underline{\theta} \otimes \hat{\underline{F}}_1(\hat{\underline{u}}(\underline{\theta})) = \underline{\theta} \otimes \underline{F}'_0(\underline{v} \otimes \hat{\underline{u}}(\underline{\theta}))/\underline{F}'_0(\underline{v}) \quad ,$$

which upon the substitution of $\underline{F}'_0(\underline{v})$ from Eq. (C.13) becomes

$$\hat{\underline{u}}(\underline{\theta}) = \underline{\theta} \otimes \underline{F}'_0(\underline{v} \otimes \hat{\underline{u}}(\underline{\theta}))/\underline{v} \otimes \underline{F}'_0(\underline{1}) \quad .$$

Because division and direct product of vectors are defined as inverse operations, this can be rewritten as

$$\underline{v} \otimes \hat{\underline{u}}(\underline{\theta}) = \underline{\theta} \otimes \underline{F}'_0(\underline{v} \otimes \hat{\underline{u}}(\underline{\theta}))/\underline{F}'_0(\underline{1}) \quad . \tag{C.14}$$

Finally, by the definition of $\underline{u}(\underline{\theta})$ in Eq. (3.19) and the universal consistency relations, Eq. (3.9),

$$\underline{u}(\underline{\theta}) = \underline{\theta} \otimes \underline{F}'_0(\underline{u}(\underline{\theta}))/\underline{F}'_0(\underline{1}) \quad . \tag{C.15}$$

Comparing Eqs. (C.14) and (C.15), one sees that the components of $\underline{u}(\underline{\theta})$ and $\underline{v} \otimes \hat{\underline{u}}(\underline{\theta})$ are solutions to the same equations and hence equal, completing our demonstration.

LIST OF SYMBOLS

A	number (or concentration) of antibodies in the system
C	concentration of antigen free in solution
C_i	concentration of antigen bound by i sites to cell surface receptors
C_0	total concentration of antigen (free plus bound)
$C(\theta^n)$	coefficient of θ^n
$E(X)$	expected value of the random variable X
$E(X\|Y)$	expected value of the random variable X conditional on the value of the random variable Y
E_{ij}	the event that a rooted tree contains i antibodies and j antigens
E_A	the event that a tree has an antibody root
E_F	the event that a rooted tree is finite
f	antigen (effective) valence
$F_0(\theta)$	probability generating function for the number of offspring from a parent in the zeroth generation
$F_1(\theta)$	probability generating function for the number of offspring from a parent in the first or later generation
$F'(\theta)$	$dF/d\theta$
$F^n(\theta)$	$F(\theta)$ raised to the nth power
$F_{A0}(\theta)(F_{G0}(\theta))$	probability generating function for the number of offspring from an antibody (antigen) parent in the zeroth generation
$F_{A1}(\theta)(F_{G1}(\theta))$	probability generating function for the number of offspring from an antibody (antigen) parent in the first or later generation
g	antibody (receptor) valence
G	number (or concentration) of antigens in the system
k_f	forward rate constant
k_r	reverse rate constant

k_i	forward rate constant describing the formation of an antigen bound to a cell at i sites, given that it is already bound at i-1 sites, i = 1, 2, ..., f
k_{-i}	reverse rate constant describing the dissociation of one bound site on an antigen bound at i sites, i = 1, 2, ..., f
K	equilibrium constant for the binding of an antigen molecule to a single antibody (receptor) site, vk_1/k_{-1}
m	mean number of sites bound on (or mean number of offspring of) an f-valent particle in the zeroth generation
$\tilde{m}$	mean number of offspring of an f-valent particle in the first or later generation
m_{ij}	concentration of aggregates containing i antibodies and j antigens
m_{is}	concentration of aggregates containing i bivalent antibodies and s singly bound bivalent antigens
$M_A(M_G)$	mean number of sites bound on, or mean number of offspring of, an antibody (antigen) in the zeroth generation
$\tilde{M}_A(\tilde{M}_G)$	mean number of offspring of an antibody (antigen) in the first or later generation
$\hat{M}_A(\hat{M}_G)$	mean number of sites bound on an antibody (antigen) in the zeroth generation, conditional on the aggregate containing the antibody (antigen) belonging to the sol
$MW_A(MW_G)$	molecular weight of antibody (antigen)
N_n	number (or concentration) of n-mers in the systems
p	probability that a site on a particle with f equally reactive sites is bound, i.e., the extent of reaction
p_k	probability that k sites are bound on an f-valent particle in the zeroth generation
$\tilde{p}_k$	probability that an individual in the first or later generation has k offspring
$p_A(p_G)$	probability that an antibody (antigen) site is bound, assuming sites are equally reactive

$p_{Ak}(p_{Gk})$	probability that k sites are bound on an antibody (antigen) in the zeroth generation
$\hat{p}_{Ak}(\hat{p}_{Gk})$	probability that k sites are bound on an antibody (antigen) in the zeroth generation, conditional on the aggregate containing the antibody (antigen) belonging to the sol
$P_{ab}^{(A)}(P_{ab}^{(G)})$	probability that a antibodies (antigens), all in the first or all in some later generation, have a total of b antigen (antibody) offspring in the next generation
q	probability of extinction of a family tree
r	ratio of total antigen sites to total antibody sites, fG/2A
s	number of singly bound bivalent antigens in a linear antigen-antibody aggregate
S	concentration of free receptor sites
S_0	total concentration of receptor sites (free plus bound)
t	time
u, u_A, u_G	dummy variables in probability generating functions
$u(\theta)$	weight fraction generating function for subtrees beginning in the first or later generation
$u_A(\underline{\theta})(u_G(\underline{\theta}))$	weight fraction generating function for subtrees beginning with an antibody (antigen) in the first or later generation
v	total number of identical determinants on an antigen i.e., its valence
$v_A(v_G)$	probability that a tree with an antibody (antigen) root is finite
w_n	weight fraction of aggregates containing n f-valent particles
w_{ij}	weight fraction of aggregates containing i antibodies and j antigens
w_{is}	weight fraction of aggregates containing i bivalent antibodies and s singly bound bivalent antigens
$w_{ij}^{(A)}(w_{ij}^{(G)})$	weight fraction of aggregates containing i antibodies and j antigens including an antibody (antigen) root
$W(\theta)$	weight fraction generating function, $\lim_{r \to \infty} W_r(\theta)$

$\hat{W}(\theta)$	weight fraction generating function for molecules belonging to the sol
W_{ij}	proper weight fraction of (i,j)-mers, defined by Eq. (3.36)
W_g	weight fraction of the gel
$W_r(\theta)$	generating function for Y_r
W_s	weight fraction of the sol = $W(1)$
$W_A(\underline{\theta})(W_G(\underline{\theta}))$	weight fraction generating function for aggregates with an antibody (antigen) root
x_ℓ	ℓth component of a vector $\underline{x}$ that satisfies Eqs. (4.32) - (4.34)
X	random variable
X_{ij}	family of f-dimensional vectors $\underline{x}$, see Lemma 4.6
y_ℓ	ℓth component of a vector $\underline{y}$ that satisfies Eqs. (4.36) - (4.38)
Y	total number of individuals in a branching process, $\lim_{r\to\infty} Y_r$
Y_{ijk}	family of f-dimensional vectors $\underline{y}$, see Lemma 4.7
Y_r	total number of individuals in a branching process up to and including generation r
Z_r	number of individuals in generation r
Z_{ir}	number of individuals of type i, i = A, G, in generation r

Greek Letters

α	$p_A p_G$
β	$\kappa S_0/2$
γ	$v_G^{1/2}$
δ_{ij}	Kronecker delta, $\delta_{ij} = 0$ if $i \neq j$, $\delta_{ij} = 1$ if $i = j$.
κ	geometric mean of cross-linking equilibrium constants
ρ	probability that an antibody is the root of a tree
θ_A, θ_G	dummy variables in probability generating functions

Superscripts

^   denotes probability or probability generating function conditional on the aggregate belonging to the sol

~   denotes probability applicable to the first and subsequent generations only

'   denotes differentiation

*   denotes the value of a variable at the gel point

n   denotes a generating function raised to the nth power

BIBLIOGRAPHY

Abramowitz, M. and Stegun, I. A. Handbook of Mathematical Functions. National Bureau of Standards, Washington, DC, 1964.

Aris, R. and Gavalas, G. R. On the theory of reactions in continuous mixtures. Phil. Trans. Roy. Soc. London A260, 351-393 (1965).

Barisas, B. G. Photobleaching recovery studies of the mobility of polymeric antigens on B cell surfaces. In: Cell Surface Dynamics: Concepts and Models, A. S. Perelson, C. DeLisi, and F. W. Wiegel, eds., pp. 167-202. New York: Marcel Dekker, 1984.

Bell, G. I. Model for the binding of multivalent antigen to cells. Nature 248, 430-431 (1974).

Bell, G. I. B lymphocyte activation and lattice formation. Transplant. Rev. 23, 23-36 (1975).

Blum, J. J. Hormone/receptor/effector interactions. In: The Receptors, Vol. 2, P. M. Conn, ed. New York: Academic Press (in press).

Bromwich, T. J. I'A. An Introduction to the Theory of Infinite Series, 2nd ed., rev. London: Macmillan, 1947.

Cohen, R. J. and Benedek, G. B. Equilibrium and kinetic theory of polymerization and the sol-gel transition. J. Phys. Chem. 86, 3696-3714 (1982).

Crothers, D. M. and Metzger, H. The influence of polyvalency on the binding properties of antibodies. Immunochem. 9, 341-357 (1972).

DeLisi, C. Physical chemical and biological implications of receptor clustering. In: Physical Chemical Aspects of Cell Surface Events in Cellular Regulation, C. DeLisi and R. Blumenthal, eds., pp. 261-285. New York: Elsevier North-Holland, 1979.

DeLisi, C. Theory of clustering of cell surface receptors by ligands of arbitrary valence: Dependence of dose response patterns on a coarse cluster characteristic. Math. Biosciences 52, 159-184 (1980).

DeLisi, C. The magnitude of signal amplification by ligand-induced receptor clustering. Nature 289, 322-323 (1981).

DeLisi, C. and Marchetti, F. A theory of measurement error and its implications for spatial and temporal gradient sensing during chemotaxis. II. The effects of non-equilibrated ligand binding. Cell Biophys. 5, 237-253 (1983).

DeLisi, C., Marchetti, F., and Del Grosso, G. A theory of measurement error and its implications for spatial and temporal gradient sensing during chemotaxis. Cell Biophys. 4, 211-229 (1982).

DeLisi, C. and Perelson, A. The kinetics of aggregation phenomena. I. Minimal models for patch formation on lymphocyte membranes. J. Theoret. Biol. 62, 159-210 (1976).

DeLisi, C. and Thakur, A. K. Antigen binding to receptors on immunocompetent cells. II. Thermodynamic and biological implications of the receptor crosslinking requirement for B cell activation. Cell. Immunol. 28, 416-426 (1977).

Dongen, P. G. J. van and Ernst, M. H. Pre- and post-gel size distributions in (ir)reversible polymerization. J. Phys. A: Math. Gen. 16, L327-L332 (1983).

Dongen, P. G. J. van and Ernst, M. H. Kinetics of reversible polymerization. J. Stat. Phys. 37, 301-324 (1984).

Dembo, M. and Goldstein, B. Theory of equilibrium binding of symmetric bivalent haptens to cell surface antibody: Application to histamine release from basophils. J. Immunol. 121, 345-353 (1978).

Dembo, M., Goldstein, B., Sobotka, A. K., and Lichtenstein, L. M. Histamine release due to bivalent penicilloyl haptens: Control by the number of crosslinked IgE antibodies on the basophil plasma membrane. J. Immunol. 121, 354-358 (1978).

Dembo, M., Goldstein, B., Sobotka, A. K., and Lichtenstein, L. M. Histamine release due to bivalent penicilloyl haptens: The relation of activation and desensitization of basophils to dynamic aspects of ligand binding to cell surface antibody. J. Immunol. 122, 518-528 (1979).

Dintzis, H. M., Dintzis, R. Z., and Vogelstein, B. Molecular determinants of immunogenicity: The immunon model of immune response. Proc. Natl. Acad. Sci. USA 73, 3671-3675 (1976).

Dintzis, R. Z., Middleton, M. H., and Dintzis, H. M. Studies on the immunogenicity of T-independent antigens. J. Immunol. 131, 2196-2203 (1983).

Dintzis, R. Z., Vogelstein, B., and Dintzis, H. M. Specific cellular stimulation in the primary immune response: Experimental tests of a quantized model. Proc. Natl. Acad. Sci. USA 79, 884-888 (1982).

Dobson, G. R. and Gordon, M. Configurational statistics of highly branched polymer systems. J. Chem. Phys. 41, 2389-2398 (1964).

Donoghue, E. Analytic solutions of gelation theory for finite, closed systems. J. Chem. Phys. 77, 4236-4246 (1982).

Donoghue, E. Relationship between the kinetic and statistical approaches to f-valent polycondensation. In: Kinetics of Aggregation and Gelation, F. Family and D. P. Landau, eds., pp. 221-224. Amsterdam: North Holland, 1984.

Donoghue, E. and Gibbs, J. H. Mean chain length distributions in finite polycondensing systems. J. Polymer Sci: Polymer Symp. 63, 131-145 (1978).

Donoghue, E. and Gibbs, J. H. Mean molecular size distributions and the sol-gel transition in finite, polycondensing systems. J. Chem. Phys. 70, 2346-2356 (1979).

Dower, S. K., DeLisi, C., Titus, J. A., and Segal, D. M. Mechanism of binding of multivalent immune complexes to Fc receptors. I. Equilibrium binding. Biochemistry 20, 6326-6334 (1981a).

Dower, S. K., Titus, J. A., DeLisi, C., and Segal, D. M. Mechanism of binding of multivalent immune complexes to Fc receptors. 2. Kinetics of binding. Biochem. 20, 6335-6340 (1981b).

Dower, S. K., Titus, J. A., and Segal, D. M. The binding of multivalent ligands to cell-surface receptors. In: Cell Surface Dynamics: Concepts and Models, A. S. Perelson, C. DeLisi, and F. W. Wiegel, eds., pp. 277-328. New York: Marcel Dekker, 1984.

Drake, R. A general mathematical survey of the coagulation equation. In: <u>Topics in Current Aerosol Research</u>, Vol. 3, pt. 2, G. M. Hidy and J. R. Brock, eds., pp. 203-376. New York: Pergamon Press, 1972.

Dusek, K. Correspondence between the theory of branching processes and the kinetic theory for random cross-linking in the post-gel stage. Polym. Bull. 1, 523-528 (1979).

Ernst, M. H., Hendriks, E. M., and Ziff, R. M. Exact solutions to the coagulation equation. Phys. Lett. 92A, 267-270 (1982).

Falk, M. and Thomas, R. E. Molecular size distribution in random polyfunctional condensation with or without ring formation: Computer simulation. Can. J. Chem. 52, 3285-3295 (1974).

Falkovitz, M. S. and Segel, L. A. Some analytic results concerning the accuracy of the continuous approximation in a polymerization problem. SIAM J. Appl. Math. 42, 542-548 (1982).

Family, F. and Landau, D. P., eds. <u>Kinetics of Aggregation and Gelation</u>. Amsterdam: North-Holland, 1984.

Feller, W. <u>An Introduction to Probability Theory and its Applications</u>, 3rd ed. New York: Wiley, 1968.

Flory, P. J. Molecular size distribution in linear condensation polymers. J. Am. Chem. Soc. 58, 1877-1885 (1936).

Flory, P. J. Molecular size distributions in three dimensional polymers. I. Gelation. J. Am. Chem Soc. 63, 3083-3090 (1941a).

Flory, P. J. Molecular size distribution in three dimensional polymers. II. Trifunctional branching units. J. Am. Chem. Soc. 63, 3091-3096 (1941b).

Flory, P. J. Molecular size distribution in three dimensional polymers. III. Tetrafunctional branching units. J. Am. Chem. Soc. 63, 3096-3100 (1941c).

Flory, P. J. <u>Principles of Polymer Chemistry</u>. Ithaca, New York: Cornell University Press, 1953.

Gandolfi, A., Giovenco, M. A., and Strom, R. Reversible binding of multivalent antigen in the control of B lymphocyte activation. J. Theoret. Biol. 74, 513-521 (1978).

Gandolfi, A., Giovenco, M. A., and Strom, R. Control of B lymphocyte activation through reversible binding of multivalent antigen: A simple model. In: <u>Systems Theory in Immunology</u>, C. Bruni, G. Doria, G. Koch, and R. Strom, eds., pp. 37-51. New York: Springer-Verlag, 1979.

Goldberg, R. J. A theory of antibody-antigen reactions. I. Theory for reactions of multivalent antigen with bivalent and univalent antibody. J. Am. Chem. Soc. $\underline{74}$, 5715-5725 (1952).

Goldberg, R. J. A theory of antibody-antigen reactions. II. Theory for reactions of multivalent antigen with multivalent antibody. J. Am. Chem Soc. $\underline{75}$, 3127-3131 (1953).

Goldstein, B. and Perelson, A. S. Equilibrium theory for the clustering of bivalent cell surface receptors by trivalent ligands: With application to histamine release from basophils. Biophys. J. $\underline{45}$, 1109-1123 (1984).

Good, I. J. The number of individuals in a cascade process. Proc. Camb. Phil. Soc. $\underline{45}$, 360-363 (1949).

Good, I. J. The joint distribution for the sizes of the generations in a cascade process. Proc. Camb. Phil. Soc. $\underline{51}$, 240-242 (1955).

Good, I. J. Generalizations to several variables of Lagrange's expansion, with applications to stochastic processes. Proc. Camb. Phil. Soc. $\underline{56}$, 367-380 (1960).

Good, I. J. Cascade theory and the molecular weight averages of the sol fraction. Proc. Roy. Soc. London $\underline{A272}$, 54-59 (1963).

Good, I. J. The generalization of Lagrange's expansion and the enumeration of trees. Proc. Camb. Phil Soc. $\underline{61}$, 499-517 (1965).

Gordon, M. Good's theory of cascade processes applied to the statistics of polymer distributions. Proc. Roy. Soc. London $\underline{A268}$, 240-259 (1962).

Gordon, M. Combinatorics and graph theory of abundance and stability of chemical species. Colloquia Mathematica Soc. János Bolyai, Vol. 4, pp. 511-523. Amsterdam: North-Holland, 1970.

Gordon, M. and Judd, M. Statistical mechanics and the critically branched state. Nature $\underline{234}$, 96-97 (1971).

Gordon, M. with Leonis, C. G. Combinatorial short-cuts to statistical weights and enumeration of chemical isomers. In: <u>Proceedings 5th British Combinatorial Conference</u>, St. J. A. Nash-Williams and J. Sheehan, eds., pp. 231-238. Winnipeg: Utilitas, 1975.

Gordon, M. and Malcolm, G. N. Configurational statistics of copolymer systems. Proc. Roy. Soc. London A295, 29-54 (1966).

Gordon, M. and Parker, T. G. The graph-like state of matter. I. Statistical effects of correlations due to substitution effects, including steric hindrance, on polymer distributions. Proc. Roy. Soc. Edinburgh A69, 181-192 (1970/71).

Gordon, M. and Ross-Murphy, S. B. The structure and properties of molecular trees and networks. Pure Appl. Chem 43, 1-26 (1975).

Gordon, M. and Scantlebury, G. R. Non-random polycondensation: statistical theory of the substitution effect. Trans. Faraday Soc. 60, 604-621 (1964).

Gordon, M. and Scantlebury, G. R. Theory of ring-chain equilibria in branched non-random polycondensation systems, with applications to $POCl_3/P_2O_5$. Proc. Roy. Soc. London A292, 380-402 (1966).

Gordon, M. and Scantlebury, G. R. Statistical kinetics of polyesterification of adipic acid with pentaerythritol or trimethylol ethane. J. Chem. Soc. London B, 1-13 (1967).

Gordon, M. and Scantlebury, G. R. The theory of branching processes and kinetically controlled ring-chain competition processes. J. Polymer Sci. C16, 3933-3942 (1968).

Gordon, M. and Temple, W. B. Chemical combinatorics. Part I. Chemical kinetics, graph theory and combinatorial entropy. J. Chem. Soc. London A5, 729-737 (1970).

Gordon, M. and Temple, W. B. Chemical combinatorics. Part 3. Stereochemical invariance law and the statistical mechanics of flexible molecules. J. Chem. Soc. London , Faraday Trans. II 69, 282-297 (1973).

Gordon, M. and Temple, W. B. The graph-like state of matter and polymer science. In: Chemical Applications of Graph Theory, A. T. Balaban, ed., pp. 300-332. New York: Academic Press, 1976.

Goursat, E. A Course in Mathematical Analysis, Vol. I, E. R. Hedrick, trans. Boston: Ginn, 1904.

Harary, F. Graph Theory. Reading, Massachusetts: Addison-Wesley, 1969.

Harris, T. E. The Theory of Branching Processes. Berlin: Springer-Verlag, 1963.

Heyde, C. C. and Seneta, E. The simple branching process, etc.; a historical note on I. J. Bienaymé. Biometrika 59, 680-683 (1972).

Hoeve, C. A. J. Molecular weight distribution of thermally polymerized triglyceride oils. II. Effect of intramolecular reaction. J. Polymer Sci. 21, 11-18 (1956).

Ishizaka, T., Conrad, D. H., Schulman, E. S., Sterk, A. R., Ko, C. G. L., and Ishizaka, K. IgE-mediated triggering signals for mediator release from human mast cells and basophils. Fed. Proc. 43, 2840-2845 (1984).

Jacobs, S., Chang, K-J., and Cuatrecasas, P. Antibodies to purified insulin receptor have insulin-like activity. Science 200, 1283-1284 (1978).

Jagers, P. Branching Processes with Biological Applications. New York: Wiley, 1975.

Jeffreys, H. and Jeffreys, B. S. Methods of Mathematical Physics, 3rd ed. Cambridge: Cambridge Univ. Press, 1972.

Kahn, C. R., Baird, K. L., Jarrett, D. B., and Flier, J. S. Direct demonstration that receptor crosslinking or aggregation is important in insulin action. Proc. Natl. Acad. Sci. USA 75, 4209-4213 (1978).

Karlin, S. and Taylor, H. M. A First Course in Stochastic Processes, 2nd ed. New York: Academic Press, 1975.

Kelly, F. P. Reversibility and Stochastic Networks, Chapter 8. New York: Wiley, 1979.

Kendall, D. G. The genealogy of genealogy. Branching processes before (and after) 1873. Bull. London Math. Soc. 7, 225-253 (1975).

King, A. C. and Cuatrecasas, P. Peptide hormone-induced receptor mobility, aggregation, and internalization. New Eng. J. Med. 305, 77-88 (1981).

Lauffenburger, D. A. Influence of external concentration fluctuations on leukocyte chemotactic orientation. Cell Biophys. 4, 177-209 (1982).

Lauffenburger, D. A. and DeLisi, C. Cell surface receptors: Physical chemistry and cellular regulation. Int. Rev. Cytol. 84, 269-302 (1983).

Liu, S.-L. and Amundson, N. R. Calculation of molecular weight distributions in polymerization. Chem. Eng. Sci. 17, 797-802 (1962).

Loor, F. Plasma membrane and cell cortex interactions in lymphocyte functions. Adv. Immunol. 30, 1-120 (1980).

Lowry, G. G. <u>Markov Chains and Monte Carlo Calculations in Polymer Science</u>. New York: Marcel Dekker, 1970.

MacGlashan, D. W., Jr., Schleimer, R. P., and Lictenstein, L. M. Qualitative differences between dimeric and trimeric stimulation of human basophils. J. Immunol. 130, 4-6 (1983).

Macken, C. A. and Perelson, A. S. Aggregation of cell surface receptors by multivalent ligands. J. Math. Biol. 14, 365-370 (1982).

Malakoff, M. and Perelson, A. S. A theory of antigen-antibody reactions in finite systems. In preparation.

Mullikin, T. W. Branching processes in neutron transport theory. In: <u>Probabilistic Methods in Applied Mathematics</u>, Vol. 1, A. Bharucha-Reid, ed. New York: Academic Press, 1968.

Peacock, J. S. and Barisas, B. G. Antigen-specific mouse lymphocyte stimulation by DNP-conjugated T-independent antigens studied by photobleaching recovery. J. Supramol. Struct. Cell. Biochem. 17, 37-49 (1981a).

Peacock, J. S. and Barisas, B. G. Photobleaching recovery studies of antigen-specific mouse lymphocyte stimulation by DNP-conjugated polymerized flagellin. J. Immunol. 127, 900-906 (1981b).

Peacock, J. S. and Barisas, B. G. Photobleaching recovery studies of T-independent antigen mobility on antibody-bearing liposomes. J. Immunol. 131, 2924-2929 (1983).

Perelson, A. S. A model for the reversible binding of bivalent antigen to cells. In: *Physical Chemical Aspects of Cell Surface Events in Cellular Regulation*, C. DeLisi and R. Blumenthal, eds., pp. 147-161. Amsterdam: Elsevier North-Holland, 1979.

Perelson, A. S. Mathematical immunology. In: *Mathematical Models in Molecular and Cellular Biology*, L. A. Segel, ed., pp. 376-403. New York: Cambridge University Press, 1980.

Perelson, A. S. Receptor clustering on a cell surface. II. Theory of receptor cross-linking by ligands bearing two chemically distinct functional groups. Math. Biosciences $\underline{49}$, 87-110 (1980).

Perelson, A. S. Receptor clustering on a cell surface. III. Theory of receptor cross-linking by multivalent ligands: Description by ligand states. Math. Biosciences $\underline{53}$, 1-39 (1981).

Perelson, A. S. A model for antibody mediated cell aggregation: Rosette formation. In: *Mathematics and Computers in Biomedical Applications*, J. Eisenfeld and C. DeLisi, eds., pp. 31-37. Amsterdam: North-Holland, 1985a.

Perelson, A. S. Paradoxes in B cell stimulation by polymeric antigen and the immunon concept. In: *Paradoxes in Immunology*, G. W. Hoffmann, J. G. Levy, and G. T. Nepon, eds. Boca Raton, Florida: CRC Press, 1985b.

Perelson, A. S. Some mathematical models of receptor clustering. In: *Cell Surface Dynamics: Concepts and Models*, A. S. Perelson, C. DeLisi, and F. W. Wiegel, eds., pp. 223-276. New York: Marcel Dekker, 1984.

Perelson, A. S. and DeLisi, C. A systematic and graphical method for generating the kinetic equations governing the growth of aggregates. J. Chem Phys. $\underline{62}$, 4053-4061 (1975).

Perelson, A. S. and DeLisi, C. Receptor clustering on a cell surface. I. Theory of receptor cross-linking by ligands bearing two chemically identical functional groups. Math. Biosciences $\underline{48}$, 71-110 (1980).

Perelson, A. S. and Goldstein, B. A new look at the equilibrium aggregate size distribution of self associating trivalent molecules. Macromolecules (in press).

Perelson, A. S., Goldstein, B., and Rocklin, S. Optimal strategies in immunology. III. The IgM-IgG switch. J. Math. Biol. 10, 209-256 (1980).

Perelson, A. S. and Macken, C. A. Kinetics of cell mediated cytotoxicity: Stochastic and deterministic models. Math. Biosciences 70, 161-194 (1984).

Perelson, A. S. and Wiegel, F. W. The equilibrium size distribution of rouleaux. Biophys. J. 37, 515-522 (1982).

Pollard, J. Mathematical Models for the Growth of Human Populations. Cambridge: Cambridge University Press, 1973.

Rembaum, A. and Dreyer, W. J. Immunomicrospheres: Reagents for cell labeling and separation. Science 208, 364-368 (1980).

Samsel, R. W. and Perelson, A. S. Kinetics of rouleau formation. I. A mass action approach with geometric features. Biophys. J. 37, 493-514 (1982).

Samsel, R. W. and Perelson, A. S. Kinetics of rouleau formation. II. Reversible reactions. Biophys. J. 45, 805-824 (1984).

Schlessinger, J. Receptor aggregation as a mechanism for transmembrane signalling: Models for hormone action. In: Physical Chemical Aspects of Cell Surface Events in Cellular Regulation, C. DeLisi and R. Blumenthal, eds., pp. 89-111. New York: Elsevier North-Holland, 1979.

Schreiber, A. B., Lax, I., Yarden, Y., Eshhar, Z., and Schlessinger, J. Monoclonal antibodies against receptor for epidermal growth factor induce early and delayed effects of epidermal growth factor. Proc. Natl. Acad. Sci. USA 78, 7535-7539 (1981).

Schreiner, G. F. and Unanue, E. R. Membrane and cytoplasmic changes in B lymphocytes induced by ligand-surface immunoglobulin interaction. Adv. Immunol. 24, 37-165 (1976).

Segel, D. M. and Stephany, D. A. The measurement of specific cell:cell interactions by dual-parameter flow cytometry. Cytometry 5, 169-181 (1984a).

Segel, D. M. and Stephany, D. A. The mechanism of intracellular aggregation. I. The kinetics of the Fcγ receptor-mediated aggregation of P388D$_1$ cells with antibody-coated lymphocytes at 4°C. J. Immunol. 132, 1924-1930 (1984b).

Seinfeld, J. H. Dynamics of aerosols. In: Dynamics and Modelling of Reactive Systems, W. E. Stewart, W. H. Ray, and C. C. Conley, eds., pp. 225-258. New York: Academic Press, 1980.

Shechter, Y., Chang, K-J., Jacabs, S., and Cuatrecasas, P. Modulation of binding and bioactivity of insulin by anti-insulin antibody: Relation to possible role of receptor self-aggregation in hormone action. Proc. Natl. Acad. Sci. USA 76, 2720-2724 (1979a).

Shechter, Y., Hernaez, L., Schlessinger, J., and Cuatrecasas, P. Local aggregation of hormone-receptor complexes is required for activation by epidermal growth factor. Nature 278, 835-838 (1979b).

Smoluchowski, M. V. Versuch einer mathematischen theorie der koagulationskinetik kolloider Lösungen. Z. Phys. Chem. 92, 129-168 (1917).

Spouge, J. L. Solutions and critical times for the monodisperse coagulation equation when $a_{ij} = A + B(i+j) + C_{ij}$. J. Phys. A: Math. Gen 16, 767-773 (1983).

Spouge, J. L. Analytic results for finite systems of ringed Flory polymers. Can. J. Chem. 62, 1262-1264 (1984).

Spouge, J. L. Polymers and random graphs: Asymptotic equivalence to branching processes. J. Stat. Phys. (in press).

Stepto, R. F. T. Intramolecular reaction and gelation in condensation or random polymerisation. In: Developments in Polymerisation - 3, R. N. Haward, ed., pp. 81-141. Barking, England: Applied Science Publishers, 1982.

Stockmayer, W. H. Theory of molecular size distributions and gel formation in branched-chain polymers. J. Chem. Phys. 11, 45-55 (1943).

Stockmayer, W. H. Theory of molecular size distribution and gel formation in branched polymers. II. General cross-linking. J. Chem. Phys. 12, 125-131 (1944).

Stockmayer, W. H. Molecular distributions in condensation polymers. J. Polym. Sci. 9, 69-71 (1952).

Taylor, R. B., Duffus, W. P. H., Raff, M. C., and de Petris, S. Redistribution and pinocytosis of lymphocyte surface immunoglobulin molecules induced by anti-immunoglobulin antibody. Nature (New Biol.) 223, 225-229 (1971).

Trubnikov, B. A. Solution of the coagulation equation in the case of a bilinear coefficient of adhesion of particles. Soviet Physics-Doklady 16, 124-126 (1971).

Tutte, W. T. On elementary calculus and the Good formula. J. Combinatorial Theory B18, 97-137 (1975).

Vogelstein, G., Dintzis, R. Z., and Dintzis, H. M. Specific cellular stimulation in the primary immune response: A quantized model. Proc. Natl. Acad. Sci. USA 79, 395-399 (1982).

von Schulthess, G. K., Giglio, M., Cannell, D. S., and Benedek, G. B. Detection of agglutination reactions using anisotropic light scattering: An immunoassay of high sensitivity. Mol. Immunol. 17, 81-92 (1980).

Watson, G. S. On Goldberg's theory of the precipitin reaction. J. Immunol. 80, 182-185 (1958).

Whittaker, E. T. and Watson, G. N. *A Course of Modern Analysis*, 4th ed., reprinted New York: Cambridge University Press, 1935.

Whittle, P. Statistical processes of aggregation and polymerization. Proc. Camb. Phil. Soc. 61, 475-495 (1965a).

Whittle, P. The equilibrium statistics of a clustering process in the uncondensed phase. Proc. Roy. Soc. London A285, 501-519 (1965b).

Whittle, P. Statistics and critical points of polymerization processes. Adv. Appl. Prob. (Suppl. 1972), 199-220 (1972).

Whittle, P. Polymerization processes with intrapolymer bonding. I. One type of unit. Adv. Appl. Prob. 12, 94-115 (1980a).

Whittle, P. Polymerization processes with intrapolymer bonding. II. Stratified processes. Adv. Appl. Prob. 12, 116-134 (1980b).

Whittle, P. Polymerization processes with intrapolymer bonding. III. Several types of unit. Adv. Appl. Prob. 12, 135-153 (1980c).

Wofsy, C. Analysis of a molecular signal for cell function in allergic reactions. Math. Biosciences 49, 69-86 (1980).

Wofsy, C., Goldstein, B., and Dembo, M. Theory of equilibrium binding of asymmetric bivalent haptens to cell surface antibody: Application to histamine release from basophils. J. Immunol. 121, 593-601 (1978).

Zeman, R. and Amundson, N. R. Continuous models for polymerization. A.I.Ch.E.J. 9, 297-302 (1963).

Ziff, R. M. Kinetics of polymerization. J. Stat. Phys. 23, 241-263 (1980).

Ziff, R. M. Aggregation kinetics via Smoluchowski's equation. In: Kinetics of Aggregation and Gelation, F. Family and D. P. Landau, eds., pp. 191-200. Amsterdam: North-Holland, 1984.

Ziff, R. M. and Stell, G. Kinetics of polymer gelation. J. Chem. Phys. 73, 3492-3499 (1980).

Biomathematics

Managing Editor: S. A. Levin

Volume 9
W. J. Ewens

Mathematical Population Genetics

1979. 4 figures, 17 tables. XII, 325 pages.
ISBN 3-540-09577-2

This graduate level monograph considers the mathematical theory of population genetics, emphasizing aspects relevant to evolutionary studies. It contains a definitive and comprehensive discussion of relevant areas with references to the essential literature. The sound presentation and excellent exposition make this book a standard for population geneticists interested in the mathematical foundations of their subject as well as for mathematicians involved with genetic ecolutionary processes.

Volume 10
A. Okubo

Diffusion and Ecological Problems: Mathematical Models

1980. 114 figures, 6 tables. XIII, 254 pages.
ISBN 3-540-09620-5

This is the first comprehensive book on mathematical models of diffusion in an ecological context. Directed towards applied mathematicians, physicists and biologists, it gives a sound, biologically oriented treatment of the mathematics and physics of diffusion.

Volume 11
B. G. Mirkin, S. N. Rodin

Graphs and Genes

Translated from the Russian by H. L. Beus
1984. 46 figures. XIV, 197 pages. ISBN 3-540-12657-0

Contents: Graphs in the analysis of gene structure. – Graphs in the analysis of gene semantics. – Graphs in the analysis of gene evolution. – Epilogue: Cryptographic problems in genetics. – Appendix: Some notions about graphs. – References. – Index of genetics terms. – Index of mathematical terms.

Springer-Verlag
Berlin
Heidelberg
New York
Tokyo

Journal of Mathematical Biology

ISSN 0303-6812　　　　　　　　　Title No. 285

Editorial Board:
H. T. Banks, Providence, RI; **J. D. Cowan**, Chicago, IL;
J. Gani, Lexington, KY; **K. P. Hadeler** (Managing Editor),
Tübingen; **F. C. Hoppensteadt**, Salt Lake City, UT;
S. A. Levin (Managing Editor), Ithaca, NY; **D. Ludwig**,
Vancouver; **L. J. D. Murray**, Oxford, **L. T. Nagylaki**,
Chicago, IL; **L. A. Segel**, Rehovot
in cooperation with a distinguished advisory board.

For mathematicians and biologists working in a wide spectrum of fields, the **Journal of Mathematical Biology** publishes:
- papers in which mathematics in used to better understand biological phenomena
- mathematical papers inspired by biological research and
- papers which yield new experimental data bearing on mathematical models.

Contributions also discuss related areas of medicine, chemistry, and physics.

Articles from a recent issue:

E. Doedel: The computer-aided bifurcation analysis of predator-prey models
S. Karlin, S. Lessard: On the optimal sex-ratio: A stability analysis based on a characterization for one-locus multiallele viability models
J. M. Mahaffy, C. V. Pao: Models of genetic control by repression with time delays and spatial effects
P. Creegan, R. Lui: Some remarks about the wave speed and traveling wave solutiions of a nonlinear integral operator
H. Aargaard-Hansen, G. F. Yeo: A stochastic discrete generation birth, continuous death population growth model and its approximate solution
F. M. Hoppe: Pólya-like urns and the Ewens' sampling formula
M. Weiss: A note on the rôle of generalized inverse Gaussian distributions of circulatory transit times in pharmacokinetics
R. Dal Passo, P. de Mottoni: Aggregative effects for a reaction-advection equation.

Subscription information and sample copy upon request

Springer-Verlag
Berlin
Heidelberg
New York
Tokyo

If you have any concerns about our products,
you can contact us on
ProductSafety@springernature.com

In case Publisher is established outside the EU,
the EU authorized representative is:
**Springer Nature Customer Service Center GmbH
Europaplatz 3, 69115 Heidelberg, Germany**

Printed by Libri Plureos GmbH
in Hamburg, Germany